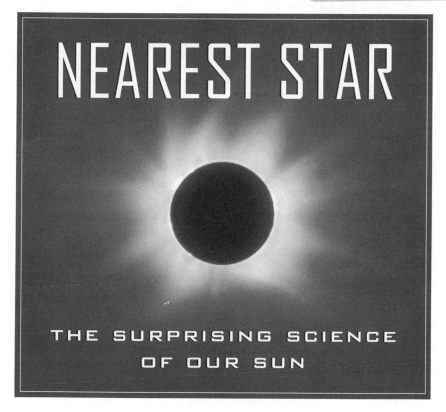

NEAREST STAR

THE SURPRISING SCIENCE OF OUR SUN

Leon Golub & Jay M. Pasachoff

HARVARD UNIVERSITY PRESS

CAMBRIDGE, MASSACHUSETTS

LONDON, ENGLAND

Copyright © 2001 by the President and Fellows of Harvard College
All rights reserved
Printed in the United States of America
Second printing, 2002

First Harvard University Press paperback edition, 2002

Title page illustration: the 1999 solar eclipse, a composite image
to emphasize the structure of the coronal streamers
(© 1999 by Jay M. Pasachoff and Wendy Carlos)

Library of Congress Cataloging-in-Publication Data

Golub, L. (Leon)
Nearest star : the surprising science of our sun /
Leon Golub and Jay M. Pasachoff.
p. cm.
Includes bibliographical references and index.
ISBN 0-674-00467-1 (cloth)
ISBN 0-674-01006-X (pbk.)
1. Sun. I. Pasachoff, Jay M. II. Title.

QB521.G65 2001
523.7—dc21
00-063213

NEAREST STAR

To the future generations, for whom we hold the Earth in trust

Contents

Preface

Our Sun is a fairly ordinary star, a bit brighter than most but not exceptionally so. There are many stars much bigger and brighter, while most stars are smaller and fainter. The Sun is not an especially variable or active star, and it has no enormous chemical or magnetic peculiarities. It is not a very young star, nor is it old and nearing the end of its life. It is, in short, truly exceptional in only one way: it is very close to the Earth—in fact, at just the right distance to make life as we know it possible.

Most of us do not worship the Sun as did many in ancient civilizations, but we certainly should not take for granted the light and heat that it provides. Left to itself, the Earth would be a fantastically frigid rock at near absolute-zero temperature. If the Sun had been slightly more massive, its high temperature would have made the Earth's surface hot enough to melt lead. A smaller Sun would have left the Earth unbearably cold and possibly subject to high levels of radiation, since smaller stars tend to have higher levels of activity, giving off devastating ultraviolet and x-rays. Distance also matters. Had the Earth been closer, we might be as infernally hot as Venus; farther away and we might have been as cold and arid as Mars. We are in the position of Goldilocks, living at just the right distance from a just-right star.

Does this mean that the planet Earth is unique and that we live in a providential "best of all possible worlds"? There are dangers with this way of thinking, flattering as it is to human sensibilities, because it inadvertently fosters a certain complacency. Since indeed other planets in our solar system do not so far appear to support life, the implication is that life requires some fairly unlikely conditions in order to flourish. Yet granted that the probability of finding Earth-like conditions is small, the number of planets in the Universe is very large (probably billions in our galaxy alone). This obviously increases the statistical likelihood of habitable planets. On this view, the Earth is not so much providentially unique as merely *rare*.

This in turn implies certain responsibilities for its inhabitants. Since life as we know it appears to be possible within only a narrow range of conditions, it would be prudent to know as much as we can about the star that provides the bedrock conditions on which our existence is founded. Moreover, our newfound ability to alter the Earth's state on a global scale brings this need into sharp focus. For example, it is not enough for the Earth to be at the right distance from the Sun, and reflect back the right percentage of the solar light it receives. The Earth's atmosphere is also of major importance in determining the global temperature. Without it, the Earth would be colder by about 33°C (roughly 60°F), and therefore a frozen lump of ice. Right now, we are making small but significant changes to the composition of our atmosphere that may, within a short time, be large enough to produce major unpleasant effects. Do the natural variations in the Sun's brightness enhance or diminish these man-made effects? How do changes in solar activity affect the formation of ozone and atmospheric circulation and weather patterns?

This book explores the Sun in a comprehensive way for the nonscientific reader who wants to gain a general idea of the range and significance of solar physics. We will explain what is known about the Sun and how this knowledge is acquired, discuss the origin of the Sun's light and heat, and explore how the Sun evolved and what it will eventually become. We will pay special attention to cutting-edge research on the Sun's outer atmosphere—the part that we can see—and the effects of this solar atmosphere on the Earth and the space around Earth. Unlike other stars, which are mere points in the sky, the Sun is so close that we can see its surface. We see sunspots form and gigantic explosive events erupt out toward the Earth. Thanks to careful measurements of the Sun's surface motions, we have recently even learned to "see" inside the Sun.

Our book continues a fine tradition at Harvard University Press of descriptive books about the Sun for general audiences. We are proud to be following in the footsteps of Donald H. Menzel's *Our Sun*, with its first edition in 1949 and its second edition in 1959. We are also proud to be in the tradition of Robert W. Noyes's *The Sun, Our Star* (1982). There is much that is new on and under the Sun in the intervening years, and it is a pleasure to be able to describe it here. One of us (JMP) got his start in solar astronomy from both the distinguished scientists who were just listed. Donald Menzel took him, as a Harvard first-year student, to a total solar eclipse, which they saw from an airplane over the Massachusetts coast, and introduced him to the changing solar surface as part of a freshman seminar. Robert Noyes took him, as a graduate student, to his first professional observing experiences by inviting him to spend a summer working with him and with Jacques M. Beckers at the Sacramento Peak Observatory, Sunspot, New Mexico. That work developed into his thesis on the solar chromosphere, with Noyes as advisor. Subsequently, as the Donald H. Menzel Postdoctoral Fellow at the Harvard College Observatory, he worked with Professor Menzel in running a Harvard-Smithsonian expedition to the 1970 total solar eclipse in Mexico. He also collaborated with Dr. Menzel on eclipse expeditions to Prince Edward Island, Canada, in 1972, and to Kenya in 1973. It could not have been foreseen that Dr. Menzel's Harvard freshman seminar would begin JMP on the set of 31 solar eclipse expeditions that have taken him around the world.

This book attempts to render all technical material in ordinary English. The book closest to the present volume is Ken Lang's excellent *Sun, Earth, and Sky*. The main difference between the two is that we approach the story from a different point of view. Rather than present science as a series of prestigious accomplishments, we invite the reader into an open-ended process of discovery. We try to show what motivates the questions that are being framed in solar physics, and how instrumental developments and theoretical creativity work together in a dynamic way to gain better insight into the Sun. Our aim is to introduce a wide and diverse audience to the substance and importance of solar physics without straining the reader's patience. If we succeed in doing this, our efforts will be amply rewarded.

Although we primarily address nonscientists, we hope that technophiles may also find the discussions worthwhile, as we devote considerable attention to instrumentation. For those who want to pursue some of our topics in a more technical fashion and who have access to the World Wide Web or the Internet, the following are Websites specializing in solar or solar-terrestrial matters:

- The Solar Data Analysis Center at NASA
  http://umbra.nascom.nasa.gov/images/latest.html
- The Space Environment Center and space weather
  http://www.sec.noaa.gov
- The Transition Region and Coronal Explorer Satellite
  http://vestige.lmsal.com/TRACE
- National Solar Observatory
  http://www.nso.noao.edu
- Eclipse Working Group
  http://www.totalsolareclipse.net
- Today's Space Weather
  http://www.spaceweather.com
- Amateur Astronomy Solar Site with Daily Images
  http://www.lpl.arizona.edu/~rhill/alpo/solar

Finally, despite our best efforts, there will inevitably be typos and errors. We apologize for this in advance, and plan to maintain an errata page with an updated list of Web links and corrections at:

- http://www.williams.edu/astronomy/neareststar

We encourage readers to notify us of typographical or other errors they find: lgolub@cfa.harvard.edu or jay.m.pasachoff@williams.edu.

NEAREST STAR

Introduction

About 15 billion years ago, for reasons we do not yet understand, the Universe came into existence. Matter as we know it did not exist and even the forces by which bits of matter and radiation interact with each other were different than they are today. Our knowledge of physics is good enough now for us to calculate the conditions prevailing back to an incredible 10^{-45} seconds (a decimal point followed by a string of zeros with a 1 in the 45th place) after it all started. Of course, this does not get us all the way back to zero or before (if the word "before" has a meaning in this context), but we think we can speak with a fair degree of confidence about how things proceeded thereafter.

By 0.000000000001 seconds of age, the four forces of nature that now exist—gravity, strong and weak nuclear, and electromagnetic—were in place, and by the age of several hundred seconds the Universe contained the familiar, so-called baryonic, matter that continues to exist today, the stuff of which ordinary atoms are made. A major turning point occurred at the age of about 300,000 years, when the Universe cooled enough for electrons to combine with the available nuclei, which were mostly protons and helium. At this point, atoms started to form and it suddenly became possible for photons of light to travel long distances without being absorbed. Before this time the Universe was opaque, and our best telescopes will not be able to look back beyond this era.

Some time later, perhaps at an age of 1 billion years, galaxies started to form. Until then, there were few stars and therefore no sources of light—the Universe was in a dark age. Since galaxies consist of large numbers of stars, many billions of stars must have been forming, and we must assume that some fraction of them had planets as well. The Universe since then has changed only in some details—galaxies have evolved, the fraction of matter in heavy elements has increased a bit— but has otherwise looked pretty much the same as it does now.

The formation of the Sun is one extremely minute part of this history, the story of one tiny star among the trillions that have come and gone during the past 15 billion years. It is a relatively young star, only 5 billion years old and thus not of the first generation. This means that it, and the planets around it, contain heavier elements formed when earlier stars became novas and supernovas. These heavier elements—oxygen, silicon, iron, carbon, and so on—make possible certain side effects, such as organic life.

The Sun is by far the brightest object in our sky, and the difference between its presence or absence overhead is literally like night and day. It is clearly far away, although it took centuries to figure out just how far. How is it, then, that we can know anything about an object that is far away, extremely hot, and astoundingly large?

The answer is that the information is in the light. The science of spectroscopy allows us to analyze the solar light in detail (see Figure 1.1 for an example) and thereby learn about the elements that compose the Sun and their physical states. If we then also use high-resolution images of the Sun, we are able to find out what physical processes are occurring to produce the type of light that we see. We can even, with the help of a new method of measurement known as helioseismology, study the interior of the Sun as well.

FIGURE 1.1   Josef Fraunhofer's photospheric spectrum, published in 1817. This was the first spectrum to show sharp, dark, absorption lines in the solar light. The locations and strengths of these numerous "spectral lines" provide information about the physical conditions on the Sun.

There is more than light coming from the Sun. Extremely high energy particles from large solar eruptions sometimes reach the surface of the Earth and are detected by terrestrial monitoring equipment. By putting instruments into space, we can extend the range of wavelengths available, enabling us to see solar phenomena not visible from the ground, and we can also intercept and study some of the actual solar material as it flows past the Earth at hundreds of miles per second.

Despite a rapid accumulation of new knowledge in the past 20 years, there is still a great deal we don't know about the Sun. The gaps in our knowledge have broad implications, because solar studies are relevant to almost all of astrophysics: many of the more exotic aspects of astronomy concerning distant stars and galaxies must, of necessity, be based on a foundation of theories and models developed and tested in the solar context. Our ability to explore the unfamiliar territory of intergalactic space reflects how well we understand the more familiar object close to home.

Solar physicists do not normally spend their days asking each other the big questions. But a group of them attending a conference might, after dinner and some wine, ask each other:

- Do we really know the energy source of the Sun? One of the great triumphs of astrophysics in the first half of the twentieth century was the understanding that the Sun is powered by thermonuclear fusion—essentially the same process that makes a hydrogen bomb, but on an enormously larger scale. The nuclear reactions ought to release a certain number of sub-atomic particles called neutrinos; these are indeed detected, but in far smaller numbers than they ought to be. Where does the problem lie? Is our understanding of nuclear physics wrong? Are our models of the interior of the Sun wrong? Is there some oddity about neutrinos that we don't yet know?

- Why is the Sun not just a bland, featureless, glowing ball? That is, why does the Sun have sunspots, flares, and a corona? Why does the activity come and go in an 11-year cycle? We now believe that all of these effects are caused by strong magnetic fields generated deep inside the Sun. Do we understand how and why the Sun generates magnetic fields? How prevalent are magnetic fields on other stars, on planets, throughout our galaxy and other galaxies? How do magnetic fields affect the formation of these? What role do fields play in more exotic objects, such as black holes, pulsars, astrophysical jets, and quasars?

· Is the Sun changing in ways that affect the Earth's global climate? Does the sunspot cycle produce climate changes on Earth? Are there longer-term changes in the Sun that might have produced the ups and downs of climate that we find in geological records? More important, will changes in the Sun in the future produce climatic effects on the Earth? Will man-made effects dominate over the solar changes, or will they reinforce each other to produce a major climate disaster?

THE SUN'S PHYSICAL PARAMETERS

Questions such as these, which fuel the excitement of solar research, require a long apprenticeship in learning what has been established so far about the Sun's basic parameters. This chapter and the next two will give the reader a taste of this process by reviewing the fundamentals.

Here are some of the basic facts:

· The ratio of the Sun's diameter to that of the Earth is: 109.
· The ratio of the Sun's mass to that of the Earth is: 333,000.
· The ratio of average solar density to that of the Earth is: 1/4.
· The ratio of the Sun's mass to the sum of all the masses of all the planets is: 744.

What do these numbers mean? The Sun is *big* by Earth standards, over a hundred times the diameter, meaning more than a million times the volume. The smallest features that we can see on the Sun with the naked eye or with low-power telescopes, such as sunspots, are usually as big as the Earth.

The Sun is also very massive, having over 300,000 times more total matter than the Earth. Since the Sun is a million times bigger than the Earth, if it had the same density as the Earth, it would be a million times more massive. But its density is low, only one fourth that of the Earth (giving it about the same density as water). The implication of the low density is that the Sun is not made of the same stuff as is the Earth. It is mainly hydrogen, the lightest element, followed by helium, the second lightest. (This fact was first realized by Cecilia Payne in her 1925 Radcliffe Ph.D. dissertation, but it was so contrary to expectation that, under pressure, she labeled her result "spurious.")

The Earth is made mostly of heavier elements, with very little hydrogen or helium, even though modern cosmology tells us that these two light elements are the most plentiful by far in the entire universe. It would seem that during the formation of the solar system something caused planets like Earth to end up with more heavy elements, or with fewer lighter elements. Today's explanation is that the smaller planets such as the Earth did not have enough gravitational pull to hold onto very much hydrogen; it escaped back into space and we ended up mainly with the relatively rare heavy elements—oxygen, silicon, magnesium, and iron being the most abundant. The large planets in the solar system, such as Jupiter and Saturn, retained these lighter elements and have far lower density than the small inner "rocky" planets. In the Sun, hydrogen and helium together make up 98 per cent of the total mass.

The fourth datum on the list explains why the Sun is the center of our solar system: it has over 700 times as much mass as all of the solar system planets combined, including comets and asteroids. All of these objects form a self-gravitating system. Floating freely in space, they are held together by their mutual gravitational pulls and are relatively uninfluenced by other distant masses. In such a system, if one of the masses is much larger than the others, it will be nearly unmoved by the gravitational pull that the other bodies exert on it. For our solar system, the Sun has about 99.9 per cent of the total mass. This means that in the gravitational tug-of-war among all these bodies orbiting around each other, it is a very good approximation—to an accuracy of about 0.1 per cent—to say that the Sun remains stationary at the center of the solar system and all of the planets, asteroids, comets, and so on, orbit around it.

THE DISTANCE TO THE SUN

As a warm-up exercise, we start by trying to figure out just how bright the Sun really is. We can get some idea by making measurements here on Earth, if we measure the brightness here and if we also figure out how far away the Sun is. The technique is this: Assume that the Sun radiates equally in all directions, so that our local data are representative of what anyone, anywhere, at our distance from the Sun would measure. This distance defines a spherical surface enclosing the Sun and having a radius equal to 1 A.U. (Astronomical Unit), the average distance between the Earth and the Sun. If we then take our measured value and multiply it by the surface area of this enclosing sphere, we have determined the total power emitted by the Sun.

In order to make this calculation, we need to know the radius of the sphere, that is, the distance between the Earth and the Sun. The problem is that no direct measurement is possible. What we can do from our terrestrial location is to measure angles between objects and use these data to figure out *relative* distances. By measuring angles we can lay out the geometric pattern of objects in the solar system and determine, for instance, that the Sun is 400 times as far away as the Moon. So if we know the distance to either one of them, we know the distance to both.

The first person known to have made such a measurement was Aristarchus in the third century b.c.e., who measured the angle between the Sun and Moon when the Moon was exactly half full. If the Sun were infinitely far away, this angle would be 90 degrees; with the Sun at a finite distance, the angle is slightly smaller, as shown in Figure 1.2. With this method, the number calculated for the distance of the Sun is extremely sensitive to the measured value of the angle. Aristarchus

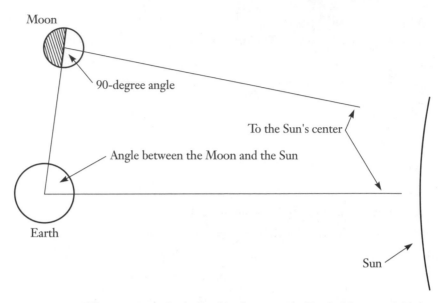

FIGURE 1.2   The geometrical relationships between the Earth, Moon, and Sun used by Aristarchus to determine the distance to the Sun.

measured 87 degrees, whereas the value is really more like 89.85 degrees. His calculation said that the Sun is only 18 times farther away from the Earth than is the Moon, rather than 400 times farther. Still, the method he used was sound, and with more accurate measurements of this sort we can determine the relative distances of the Sun, Moon, and planets quite well.

But we still do not know the scale of this pattern of relationships. How do we ever find any absolute distances? The answer to this problem turned out to require the invention of extremely accurate clocks. Why clocks? Because the method used was triangulation from widely separated points on the Earth, and the method requires not just a measurement, but an absolute determination of when the measurement is made.

Transits of Venus

In astronomy we speak of parallax, rather than triangulation, to denote the well-known phenomenon that two objects line up differently along the line of sight for one observer than for another. For example, during a solar eclipse an observer at one location might see the Sun and Moon line up perfectly, so that the eclipse is total. But an observer some distance away might be able to see past the edge of the Moon for a partial view of the solar disk; for her, the eclipse will not be total.

We have known the relative distances among the planets for quite some time—the values have not changed much since the days of Copernicus. We have also known how fast the planets move around in their orbits, so that the angles between them and how these angles change with time has been known for many years. But in order to progress from a relative diagram to one whose absolute size is determined, we need to know the true length of any one piece of the figure; we will then know all of the lengths. Parallax can be used to make this measurement.

We do have access by direct measurement to one length: the size of the Earth, or more to the point, the distance from one side of the Earth to the other. All we need then is to line up two objects from one side, then from the other, and measure the size of the angle between these two perspectives, and we will know the absolute scale of the solar system. But what to use to make the measurement?

In 1716, Edmond Halley—who not only plotted the orbit of the comet that bears his name, but who also was the first accurately to predict the path of a total solar eclipse—pointed out that the passage of the planet Venus across the face of the Sun could be used to provide the needed marker. There had been a pair of transits of Venus in 1631 and 1639, but no useful measurements were made. The next pair of transits would occur in 1761 and 1769, and Halley, knowing he would not live to see them, urged future astronomers to make the extraordinary efforts needed to obtain the crucial measurements from widely separated parts of the Earth.

Time enters into the measurement because the contact between Venus and the bright disk of the Sun occurs at different times at the two separated sites. If we imagine that the line joining the edge of the Sun to Venus is continued out until it hits the Earth, then this line sweeps across the Earth as Venus moves across the face of the Sun. First it hits one edge of the Earth (observer number 1), then it sweeps across until it hits the other edge of the Earth (observer number 2). From our scale model—whose absolute size we are trying to determine—we know the rate at which this line is sweeping around. A measurement of the time between the transits seen at the two sites therefore tells us how much angle has been swept out in moving from one to the other. Knowing that angle and the size of the Earth, which is one side of the triangle, allows us to calculate the size of the side of the triangle formed by the distance between the Earth and Venus.

The entire business of getting to the appropriate sites, bringing accurate enough clocks, avoiding hostile natives, obtaining the data, and getting home safely was a major international undertaking, but it was accomplished, and the distance from Earth to Sun was determined.[1]

The nature of the orbits of the Earth and Venus is such that transits of Venus come in pairs with 8 years between them, and then over 100 years until the next pair. There haven't been transits of Venus since 1874 and 1882 (Figure 1.3), and the next transits will be on 8 June 2004 and on 5–6 June 2012. The 2004 transit will be visible from all of Europe, with part of it visible from the eastern United States. The whole event, from immersion of Venus onto one side of the Sun until emersion on the other side, will last over 6 hours. The following pair of transits won't occur until 11 December 2117 and 8 December 2125. None of these transits is as dramatic to see as total solar eclipses, since you have to look through a telescope to see them. But intellectually they are superb.

Transits of Mercury also occur. They are more frequent than those of Venus, coming about 13 times per century, but are very difficult to observe because Mercury is so small. The TRACE satellite (the Transi-

1. The story of the transit expeditions in the eighteenth century is told in the delightful and often hair-raising book *The Transits of Venus*, by Harry Woolf, published by Princeton University Press in 1959. See also the series of articles by Don Fernie in *The American Scientist* listed in the bibliography at the back of this book.

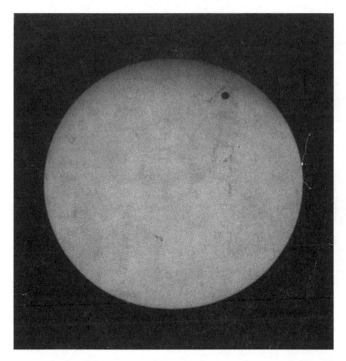

FIGURE 1.3   The transit of Venus of 1882, photographed at Vassar College by Maria Mitchell and her students.

tion Region and Coronal Explorer) observed such a transit toward the end of 1999 (Figure 1.4). The next Mercury transit will occur on 7 May 2003.

Luminosity

Determining the Sun's distance from the Earth with any reasonable accuracy was a major achievement that required much ingenuity and many centuries of work. Once the distance to the Sun is known, we can determine the Sun's total brightness, as described above: we put our detector in the sunlight, measure the amount of energy falling on the detector, and then calculate what would have been measured by a detector completely enclosing the Sun.

The detector must be one that takes into account all wavelengths, in-

FIGURE 1.4   The transit of Mercury on 15 November 1999, observed from Earth orbit by the TRACE satellite. Mercury, seen in silhouette, is the black spot at the sun's edge.

cluding the invisible infrared and ultraviolet wavelengths. Such a device is called a bolometer, and the brightness measured this way is called bolometric luminosity. It is best if we place the bolometer above the Earth's atmosphere, so that the absorption of sunlight by air is avoided. Fortunately, this can be done these days, and the solar constant (which, as we'll see, turns out to be not really a constant) can be measured to an accuracy of a small fraction of a per cent.

The number turns out to be staggeringly large by human standards: the Sun emits 380 billion billion megawatts (380 yottawatts) of power. (Note that power has dimensions of energy per unit of time; 1 watt is 1 joule per second.) In one thousandth of a second, the Sun emits enough energy to provide all of the world's current energy needs for 5,000 years. Unfortunately, we have no way at present to harness this energy, except for a minuscule fraction of it that we capture by using devices such as photovoltaic cells, solar collectors, wind farms and—indirectly—the burning of fossil fuels.

Energy from the solar interior radiates out into space from each part of the solar surface. The total power emitted by the Sun can be calculated by multiplying the intensity of the emission from each piece of its surface by the total surface area. We can also do the inverse of this calculation, in which we know how much power the Sun emits (for example, by the type of measurement described above) and we know how much surface area it has, to figure out how much power is being emitted by each piece of its surface. This in turn allows us to estimate the temperature, since the radiating power of a surface depends very strongly on its temperature. In order for the Sun, with a surface area of six billion billion square meters, to radiate away 380 yottawatts, it must have a temperature of about 6,000 K, or 10,000°F. (K is the symbol for kelvin, a unit of temperature equal in size to a Celsius degree—9/5ths the size of a Fahrenheit degree —but measured upward from absolute zero, −273°C. For temperatures as high as those in stars, we can generally ignore the 273°C offset between the Kelvin and Celsius temperature scales.)

HOW WE STUDY THE SUN

Scientific instruments extend our observing range beyond the capabilities of our senses. we make distant objects clearer with a telescope, or small objects visible with a microscope; we build instruments that can see infrared, radio, ultraviolet and x-ray wavelengths; we build particle detectors to measure cosmic rays or reaction products from the collisions of high-energy particles; and so on. The possibilities are endless, and we will give merely a small sample of the types of instruments that are used today to study the Sun.

Telescopes

Telescopes that image the Sun have special problems not encountered in night-time astronomy, mainly caused by the extreme brightness of the Sun. First, there is the well-known problem that you should never

FIGURE 1.5   The solar observatory at Big Bear Lake, California.

look at the Sun directly through the eyepiece of a telescope, since this would cause major damage to the eye. So either the solar telescope must be equipped with some type of filter that reduces the brightness of the image, or the image must be projected onto a screen for indirect viewing. Electronic recording via a CCD (charge-coupled device) camera with suitable exposure times is also a possibility.

Selection of the observing site is of major importance for solar, and indeed for all, telescopes because the amount of turbulence in the atmosphere has a major impact on the quality of the images obtained. Some of the best sites are surrounded by water, such as the solar telescopes in Hawaii, in the Canary Islands, or on an artificial island in Big Bear Lake, California, now connected to the shore by a causeway (Figure 1.5).

Once these problems are overcome, another set of problems is encountered. The light from the Sun heats the mirrors, causing them to deform and ruining the image quality. For instance, one of the standard telescope designs is a Cassegrain, in which a large mirror (the primary) focuses light onto a smaller mirror (the secondary) in order to produce a highly magnified image in a fairly compact system. A typical magnification factor for the secondary mirror in a Cassegrain is 6X, which means that the intensity of light falling on the secondary is roughly 36 times as large as that falling on the primary mirror ($36 = 6 \times 6$). For starlight this is not a problem, but for sunlight the 36 solar constants of intensity hitting the secondary mirror amount to nearly 50,000 watts per square meter. Even if the mirror has a highly reflective coating, absorption of only a few percent of this intense light will cause severe heating and distortion of the optical surface.

It also may happen that the solar image being focused by the mirrors heats the air inside the telescope tube, since those same thousands of watts are passing through the air. Air is not 100 per cent transparent, and absorption of even a small fraction of the solar energy will heat the air, which then starts to expand and convect away from the heating site, resulting in local atmospheric turbulence inside the telescope that again ruins the image. To solve this problem, which is peculiar to solar telescopes, several sites have built vacuum tower solar telescopes, in which the entire telescope interior—roughly the size of an elevator shaft—has been made into a vacuum chamber. The large solar tower in Sunspot, New Mexico, shown in Figure 1.6, is such an instrument. Note that the entire telescope is suspended on a trough of liquid mercury (labeled "Turret Mass Float" in the diagram) in order to reduce vibrations, and that the observer stands on a platform attached about two thirds of the way up the tower (labeled "40 Foot Dia. Table") in order to have access to the focused image. This facility is part of the National Solar Observatory (NSO), which also operates the McMath-Pierce Solar Telescope—the largest solar telescope in the world—at Kitt Peak near Tucson, Arizona (Figure 1.7).

In addition to data from telescopes operating in visible and infrared light, there is also much to be gained by studying the Sun at very long electromagnetic wavelengths, in the radio portion of the spectrum. The first detections of the Sun at radio wavelengths were outbursts ge-

FROM SUN
TURRET
136 FEET ABOVE GROUND
AZIMUTH MIRROR (44 IN. DIA.)
TURRET MASS FLOAT (MERCURY)
IRIS
SLIP RINGS
COOLING PIPES
FLOAT (CAPACITY, 250 TONS)
FLOAT SUPPORT RODS
VIBRATION ISOLATION SYSTEM
CENTERING BEARING
HELIOSTAT
GUIDER OPTICS
PAINTED TITANIUM DIOXIDE WHITE
QUARTZ WINDOW (24 INCH DIA.)
GUIDER
40 FOOT DIA. TABLE
10,000 FT·LB. TORQUE MOTOR
VACUUM TUBES ROTATE (250 TONS)
0.0222 RADIAN OFF AXIS
UPPER FOCUS EXTREME
SPHERICAL MAIN MIRROR (64 IN. DIA., 180 FT. FOCAL LENGTH)
LOWER FOCUS EXTREME
VACUUM SWIVEL MERCURY SEAL)
CENTERING BEARING
SUMP PUMP

QUARTZ WINDOW (30 IN. DIA.)
ELEVATION MIRROR (44 IN. DIA.)
SWING DOOR
ELEVATOR (TOP POSITION)
ELEVATOR DOOR
ELEVATOR HOIST
CONCRETE TOWER
LADDER
3 FT. THICK CONCRETE
AZIMUTH SEAL (MERCURY)
ELEVATOR RAILS
HOIST
SPECTROGRAPHS (3) (EVACUATED, 5 FT. DIA. 68 FT. LONG)
UPPER TUBE (4 FOOT DIA.)
ELEVATOR
ELEVATOR (EXTENDED)
FOCUS
GROUND LEVEL
CENTERING BEARING
LOWER TUBE (10 FOOT DIA., EVACUATED)
50 FT.
MIRROR TILTS TO SEND LIGHT TO OTHER INSTRUMENTS
12 IN. DIA. VACUUM PIPE
193 FT. BELOW GROUND
221 FT. BELOW GROUND

FIGURE 1.6   Diagram of the Richard B. Dunn Solar Telescope at the National Solar Observatory's site on Sacramento Peak, New Mexico. Note that light travels down 329 feet in a vacuum to the primary mirror 188 feet underground. The entire 200-ton rotating section that includes the main mirror, the focal region, and the spectrograph, is suspended at the top in a 10-ton tank of mercury.

FIGURE 1.7   The McMath-Pierce Solar Telescope at Kitt Peak, Arizona. The
telescope is the triangular structure in the foreground.

nerically called "noise storms" (described below) but the more normal
nonflaring Sun can now be imaged as well.

One of the problems of building a radio telescope is that, in order to
have sharp images, a mirror must be a certain size, which depends on
the wavelength of the light being imaged. This is not a problem in the
visible part of the spectrum, where the wavelength is a tiny fraction of
a hair's width, and mirrors only a few inches in diameter can begin
to give high-quality images. But radio waves are thousands of times
longer than visible-light waves—several inches to several feet between
peaks. The corresponding telescope diameter needed for imaging the
Sun would be hundreds to thousands of feet, and the instrument would
need to be pointable, since the Sun moves across the sky.

Such a device is not practical to build, so that clever alternative
methods must be used, the main one being arrays of smaller antennas
arranged so that they can be treated as if they were a single large an-
tenna. One of the first such "radioheliographs" was an east-west linear
array of 32 aerials stretching in a line nearly a mile long, constructed at

Nançay in France in the 1950s. Such an array is long in one dimension and short in the other, so that the resolution of the array is good in one dimension only. The Culgoora radioheliograph in Australia (no longer in operation) was built as a circular array laid out on a 2-mile diameter circle, thereby providing two-dimensional, rather than just one-dimensional, images.

The individual elements of such arrays must be properly spaced and the signals from the elements must be carefully combined, but the technique works and is being used for radio imaging of the Sun. Another two-dimensional arrangement is a crossed pair of linear arrays known as the Mills Cross; this arrangement was used at the Clark Lake, California, radioheliograph. The National Radio Astronomy Observatory's Very Large Array (VLA), in Socorro, New Mexico, was not built specifically as a solar instrument, but it is used during a portion of its observing time to acquire high-quality solar radio imaging observations (Figure 1.8). At present, the premier facility dedicated to solar ra-

FIGURE 1.8   The central part of the VLA, looking toward the South.

FIGURE 1.9   The Nobeyama Radioheliograph.

dio observations is the Nobeyama Solar Radio Observatory in Japan
(Figure 1.9).

Spectroscopy

Before we get to a discussion of spectroscopy and other instruments for
studying the Sun, we need to make a detour into the nature of light and
color. Historically, humans have had great difficulty in understanding
what light is. Is it something emitted by burning objects, or does the
burning activate the surrounding space in some way? Does the light fill
all of space instantaneously, or is time required? Do our eyes emit some
kind of ray that reaches out and senses an object, or does something
from an object enter our eye? If the latter, how does the shape, size, and
distance of an object register accurately on our senses? What is color?
Is it something separate from light?

   The properties of light that we find so puzzling are almost all due to
the following result of quantum mechanics: light consists of particle-
like bundles called photons, which can be created and destroyed, and

which have zero rest mass. Light does not exist "in" a fire, but is produced by the hot atoms of the fire. Light does not flow into our eyes and stay there, like water poured into a glass, but disappears—ceases to exist—when it hits the photoreceptors of the retina. Light therefore seems more ephemeral than, say, molecules of water, which can evaporate from the surface of a pond, or fall as rain and collect, but the total number of which remains constant.

Light also travels very quickly by human standards. Because of its zero rest mass, a photon of light travels (in a vacuum, that is, in the absence of matter) at the maximum possible speed, which (no great surprise) is known as the speed of light. This speed, roughly 186,000 miles per second, is so great that it might as well be infinite, as many ancient people thought, since high-precision equipment is required to make the measurement. (Galileo tried, by sending a friend to a distant mountain with a lantern and having him uncover his lantern when he saw the light from a lantern uncovered by Galileo. Good experimenter that he was, Galileo reported that the speed was too great to determine by this method, which is exactly the right conclusion.)

In addition to being so ephemeral, light also has the puzzling property of being associated with color. But we do not see color flashing past us, so it can't be light. Is it a property of objects that is revealed by light? Is light one thing, and color another? (Technically, that is the case for standard TV broadcasts, which have separate "luminance"—brightness—and "chrominance"—color—pictures.)

That white light can be spread out into colors has long been known. Isaac Newton carried out a long series of experiments in the 1660s using sunlight and prisms and showed that incoming white light will form a spectrum when passed through a good-quality prism. From the observation that the different colors are bent by different amounts in the prism (Figure 1.10), he concluded that white light is made up of "difform Rays, some of which are more refrangible than others." We now know that this "difformity" is associated with light of different wavelengths.

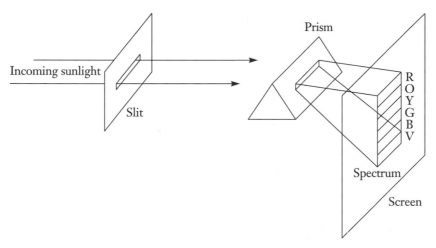

FIGURE 1.10   Using a prism to split light into its component colors.

It was William Wollaston in 1802 who commented on the "gaps" between what he saw as four colors, with some lesser gaps also indicated. Instrument-making techniques were then rapidly evolving and one of the opticians who improved the techniques was Josef Fraunhofer in Germany. He used a spectrograph with a slit, which prevented the overlapping of colors from different parts of an extended source. In 1811 he produced an epochal chart of the solar spectrum, which is a continuous band of color and shows hundreds of dark lines across it, of which he drew only the strongest, as shown in Figure 1.1. We now know that this continuous spectrum with absorption lines is from the solar photosphere. Explaining the origin of these lines led physicists, and will now lead us, to the quantum theory.

Every element has its own unique arrangement of electrons. These negatively charged electrons (thus called from the arbitrary sign convention established by Benjamin Franklin) are drawn to the positively charged nucleus according to the number of charges in that nucleus, and normally just enough electrons are attracted to make the atom as a whole neutral. For instance, carbon, with six positively charged protons in its nucleus attracts six electrons, which fall into place in a very specific arrangement around the nucleus. There is no consideration of one type of electron or another, because they are all the same.

This is a crucial point, because it means that on the atomic level, every atom of neutral carbon is *exactly* the same as any other one (except for being at a different location in space). This sort of thing does not happen in the "macro" world, the world of our ordinary experience. In this macro world, things can be "pretty much" the same, but are extremely unlikely to be truly identical. But identity at the quantum level has a bizarre (to us) consequence: when two identical bits of matter interact, they can change places, and continue on as if one were the other, and vice versa. This would be truly astounding at macroscopic scales and shows that the word "identical" has a very definite technical meaning in physics.

One consequence of quantum physics is that each type of atom has a very well defined and unique way of absorbing and emitting light, which is directly connected to its specific arrangement of electrons. The electrons around each atom can exist only in certain discrete configurations, which means that they can absorb and emit energy only in certain prescribed amounts. (Think of climbing a ladder: you can stand on discrete rungs, but there is nowhere to stand in between. Now imagine that there does not exist any in-between: you can only be on one rung or another, and nowhere else.) This form of interaction between each type of atom and the photons of light leads to a characteristic pattern of light absorbed by the element or emitted by it.

The pattern of light emitted or absorbed by an element provides a unique fingerprint for the presence of that element. A familiar example is the element sodium, which when heated emits light very strongly at a very narrow and specific range of wavelengths in the yellow part of the spectrum. The presence of small amounts of sodium in salts gives many flames a characteristic yellow color.

The Spectrograph
As shown in Figure 1.1, the light coming from the Sun can be spread out in wavelength to reveal an enormous amount of detail, in the form

of sharp variations in the strength of the light at different wavelengths. You need only glance at such a display—also known as a spectrum—to guess that there is a great deal of information contained in these variations, if you only knew how to decode it. (A commercial bar code can similarly be decoded.) Extracting information about the physical state of the object emitting the radiation from a detailed study of the spectrum became one of the main activities of astrophysicists in the nineteenth and early twentieth centuries.

An instrument used to make and record a spectrum is called a spectrograph. When an eye or a piece of film receives light, it records a range of wavelengths all at once. In a black-and-white photograph the wavelengths from red through blue are all added together, while in a color photograph some separation is made into three broad bands, such as red-yellow-blue. But in none of these cases is the discrimination fine enough to discern the very narrow (in wavelength) features, known as spectral lines, that will provide the solar physicist with enough information to, for instance, determine the chemical composition of the gas. A special instrument that disperses the light, strongly spreading out the wavelengths, is needed.

To make a spectrograph, what is required is a device that bends different wavelengths of light by different amounts. With such a device, the wavelengths can be spread out spatially so that each wavelength appears in a different location on the detector. Two main methods are used, the prism and the diffraction grating. The prism works as described above in Newton's experiments: different wavelengths are bent (refracted) more or less strongly according to their wavelength, so that a ray of white light entering the prism is split into its component colors when exiting the prism, as illustrated in Figure 1.10.

The other main method used for producing spectra is the diffraction grating. Here, a series of very closely spaced lines is cut into a surface, and the light is reflected from the grooves. When the reflected waves are in phase with each other, the reflection is enhanced; otherwise they cancel each other out. The enhancement or canceling of each wavelength occurs at a particular angle, so that each wavelength comes out of the grating at a characteristic angle. The result is thus similar—you see a spectrum of colors—to the effect of a prism, although it turns out that gratings can be made to produce a much stronger spreading of the light than prisms can.

In all likelihood, you have such a diffraction grating at home—a compact disk. If you hold its surface to the light you will see a spectrum of colors from the closely spaced rows of reflecting surfaces used to record data on the disk.

The Coronagraph

On rare occasions—roughly once every four centuries at a given spot on the Earth's surface—the Moon passes in front of the Sun so as to block out all of the Sun's visible light. The Sun is so bright that a partial eclipse, in which the Moon does not completely cover the entire disk of the Sun, might not even be noticed without advance warning. Even when 90 per cent of the Sun is covered, the amount of darkening is roughly the same as moderate cloud cover.

But when that last little bit of the Sun is covered, the effect is dramatic. The world suddenly grows quite dark and bright stars become visible in the sky. A kind of glowing aura called the corona surrounds the black disk of the Moon (Figure 1.11). It was not until the nineteenth century that it became clear that this corona actually belonged to the Sun, rather than being a result of scattering in the Earth's atmosphere, or even an effect of an atmosphere on the Moon illuminated by the bright Sun behind it.

The corona is extremely faint. It is only about one millionth as bright as the disk of the Sun, and cannot normally be seen. During an eclipse, the bright disk is covered up, and the amount of light scattered by the Earth's atmosphere is reduced enough to allow the bright inner portion of the corona to be seen. How, then, could astrophysicists see the corona at will, without waiting for an eclipse and traveling to some distant location?

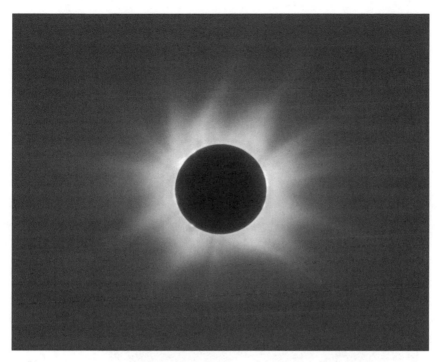

FIGURE 1.11   The 11 August 1999 total eclipse, photographed from Romania.

The enormously difficult problem of viewing the faint corona near the bright solar disk in the absence of an eclipse was partially solved in the 1930s by Bernard Lyot. First, a site with an exceptionally clear sky must be found, because humidity and air pollution or dust scatter light, causing the brightness of our atmosphere around the Sun to swamp the coronal brightness. Then a telescope is fit with a device that covers the bright solar disk; this is often done with a small circular plate, called an occulting disk, which hides the bright part from view, allowing the faint outer part, the corona, to be seen. But merely covering over the bright disk is not good enough, because the full sunlight still falls on parts of the instrument just outside of the occulting disk's shadow. The rest of the telescope must therefore be built with extreme care given to minimizing the light scattered by the lenses or mirrors, by reflections and scattering between one lens and another, and by

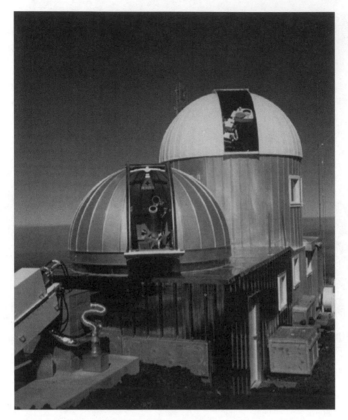

FIGURE 1.12   Left: The MK3/4 coronagraph is visible in the opening of the larger dome at the Mauna Loa Solar Observatory, Hawaii. Right: A comparison of the image quality with the old MK3 and the new MK4 is also shown.

an inevitable optical effect called diffraction. This last effect is removed by careful positioning of internal baffles (obstructions), known as Lyot stops. A good coronagraph these days will have a reduction of a factor of 1 trillion in the level of internal scattered light compared to that of the incoming light.

The sky is too bright, even viewed from high mountains, to see the corona well. At the 3,440-meter (11,290-foot) level of the Mauna Loa Solar Observatory in Hawaii, the High Altitude Observatory (HAO) runs a coronagraph that uses polarization to distinguish the corona better (Figure 1.12). Since the corona is highly polarized and the sky is not, its analysis displays the corona farther out from the sun than is visible with ordinary coronagraphs. The new MK4 corona-meter at Mauna Loa provides greatly improved coronal imaging, out to several solar radii from the limb (edge) of the Sun.

The Once and Future Sun

It would be overly optimistic of us to think that we have any accurate understanding of the conditions that make life possible. We do know that life exists on Earth, that it does not seem to have ever existed on the Moon, and that it may or may not have existed on Mars and Venus. If there is life elsewhere, such as on a moon of Jupiter, we have not yet found it. This means that we have exactly one data point, and it is dangerous to generalize from a single example.

Still, as far as we can tell, light and heat have been of central importance. The Earth would not have life on it if the Sun had not brought the surface temperature of our planet above the freezing point of water, but well below the boiling point. We are not in a good position to argue that this range of temperatures is absolutely essential for life, but it is generally necessary for the types of life that we see here. Because the presence of the Sun as a source of heat and energy is vital for life on Earth, we may be tempted to ask some pointed questions: Why is this object emitting so much light and heat? How long has it been doing this? How long will it keep doing this?

We currently have fairly reliable answers to these questions. We think that the Sun started as a collapsing cloud of gas and dust about 5 billion years ago, and that it will end up becoming a red giant, expanding in size to engulf the Earth, about 5 billion years from now. The story of how we came to our present understanding covers a wide range of fields, from geology to atomic physics, and it provides a nice illustration of the meandering and often unpredictable character of scientific progress.

THE EARLY SUN

It has been said that in the early stages of development in a field of science—before things are understood—one begins by classifying. Astronomy is no exception, and one way of classifying stars that has turned out to be highly productive is the color-magnitude diagram. Careful examination of nearby stars (where "nearby" is a relative term—the closest is 20 trillion miles away) shows that they are not all the same brightness and color, but rather come in a wide range of sizes. What is most striking, and pleasing to the orderly mind, is that there is a nice correlation between the brightness of a star—its luminosity— and its color or temperature.

Figure 2.1 is a diagram that shows the observed correlation between brightness and temperature for a sample of more than 16,000 stars. This type of display is widely used in astronomy, and is known as a Hertzsprung-Russell diagram after the two early-twentieth-century astronomers who invented it. Each dot in this figure represents a star, located horizontally on the figure from its color and vertically from its brightness. When enough stars are plotted we see that there is a large group of stars that fall in a band cutting across the diagram, from upper left to lower right. The brightest stars are the hottest and the faintest stars are the most red, or coolest. This swath is known as the "main sequence." (There is another branch containing fewer stars that heads up and to the right, which we will discuss later.)

Such a correlation demands an explanation. We might start by assuming (guessing) that this narrow band in the diagram means that there is one parameter that is varying among this assortment of stars, but that they are otherwise the same. This would mean that:

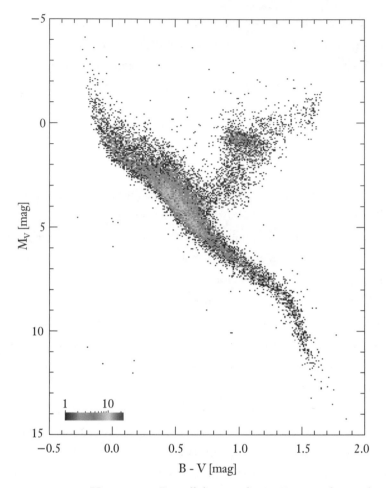

FIGURE 2.1   Hertzsprung-Russell diagram for 16,631 stars, from volume 1 of the Hipparcos and Tycho catalogues, the most accurate sets of star positions ever measured. Hipparcos (which stands for High-Precision Parallax Collecting Satellite) was a project of the European Space Agency. The Tycho catalogue covers a more extensive set of stars to a lower precision, also measured with the Hipparcos satellite.

1. All of these stars are being powered by the same process, since the pattern forms a continuous distribution across the diagram. Several different processes are very unlikely to connect together in such a smooth manner.

2. The brighter the star, the more strongly the process operates,

because the brightness of the stars in the diagram correlates with the temperature, with hotter stars being brighter.

3. These stars are all made up of roughly the same stuff, since the band is fairly narrow. Whatever the energy-release process, it seems unlikely that different fuels would all produce exactly the same correlation between size and brightness that is observed; if they did not, the band would not be so narrow.

Although it is conceivable that many separate stars will all fall into line as observed by accident, or with each one having some different power source inside it, the most economical explanation and the most reasonable is that there is one process operating inside of stars and that it is essentially the same in all of them. The essence of science, though, is not merely qualitative suggestions, but the calculation of quantities. We would therefore need to develop a theory that connects, for instance, the amount of energy generated with the mass of the star; this is our second principle above. Such a theory, in which stars are powered by nuclear fusion, has been developed quite successfully, and we will discuss it in the next section.

The third principle takes into account that the process going on inside of stars to generate energy depends on which atoms exactly are fusing together. With the growing realization early in the twentieth century that stars are made up mostly of hydrogen, it was natural to think of hydrogen nuclei—protons—as the main ingredient in this cosmic cooking.

The observations therefore imply that stars begin life by settling down to a long, slow phase of life on the main sequence: if they spent a significant amount of time doing something different, then we would see a large number of them located elsewhere on the diagram. But what happens before and after this main sequence phase? Logically enough, the earlier evolution is called the pre–main sequence phase, and it is the time when the cloud of gas and dust is contracting and heating until the center becomes hot enough to ignite nuclear fusion. The later stage, for a star like the Sun, is called the red giant phase, which forms the upper-right branch in the diagram. As the name implies, this means that the Sun will eventually go through a change that increases its size enormously, and it will swell to about the size of Mars's orbit. We have roughly 5 billion years to prepare for this disaster.

THE AGE PROBLEM

The nineteenth century saw the rise of geology as a science and continual improvement in the methods for establishing the age of the Earth. At the beginning of the century there was not much reason to doubt the age of about 6,000 years that came from calculations based on biblical references. By the middle of the century it was clear that the Earth was at least several million years old, and by the end of the century an age of a billion years seemed likely. By the start of the twentieth century, the natural radioactive decay of elements had been discovered and this eventually allowed an accurate dating of nearly 5 billion years for the age of the Earth (see Chapter 7).

Because the Sun was presumably shining during all of these billions of years, it became problematic to determine the source that could produce such an enormous amount of energy over such a long span of time. It was known that ordinary chemical processes could not keep the Sun burning more than a few thousand years—an easy calculation that merely takes the amount of mass in the Sun and assumes that it is, for example, a lump of coal that burns—so more exotic processes had to be explored.

In the 1840s, a German physician named Julius Mayer sent a paper to the Paris Academy proposing the energy released by meteors falling into the Sun as the source. There is not a huge amount of meteoritic material falling out of the sky, but the amount of energy released by each bit of matter in falling into the Sun's very strong gravitational field is large, so this mechanism was at least conceivable. A Scottish engineer in Bombay, John Waterston, submitted a paper to the Royal Society of London proposing that a contraction of the Sun under its own gravitational pull was the energy source. Again, the very large gravitational pull of an object as massive as the Sun makes this a viable mechanism as well. Both papers were rejected.

Waterston continued to work out his theory and in 1853 he gave a talk to the British Association for the Advancement of Science, in which he presented a general discussion of the formation of the solar system out of a rotating disk of gas (an idea that had been discussed by the philosopher Immanuel Kant in the previous century), and then went on to talk about the energy source of the Sun in the context of gravitational forces. Waterston included both meteor infall and gravitational contraction in the discussion, and the talk influenced two major figures in the physics world, William Thomson of Glasgow (later Lord Kelvin) and Hermann Helmholtz of Königsberg.

Thomson favored the meteor infall theory and went on to calculate, based on an estimate of the number of meteors still floating around the solar system, that the Sun could last for only another 300,000 years. Helmholtz, working within the view that the Sun had contracted out of a cloud of gas and dust, calculated that the Sun could still be contracting slowly and that this could supply the energy needed to power it for millions of years.

Both of these mechanisms turned out to be inadequate once it became clear that the Earth was billions of years old. As it happened, Helmholtz had also mentioned that the lifetime of the Sun could be longer if an altogether new source of energy were discovered, and he lived long enough to hear Ernest Rutherford proclaim, in the early years of the twentieth century, that the discovery of radioactivity provided just such an answer. It was not until the 1920s and 1930s that this new source of energy became known and was understood well enough to apply it to the problem of how the Sun is powered.

HOW THE SUN FORMED

Occasionally, in the vast spaces between the stars of our galaxy, regions of dense gas form out of the bits of debris spewed out by old, exploding stars. By interstellar standards the amount of matter in these clouds is enormous—over a millionth of an ounce per cubic mile—and if the cloud is large enough it begins to collapse in on itself because of its own gravitational pull. Small clouds, less than about a trillion miles across, do not have enough gravity of their own to collapse against the countervailing outward pull of the other stars and galactic material swirling past. But the larger clouds will continue contracting for a million years or so, and the material in the middle of the cloud will get hotter as the gas and dust are compressed by gravity.

If the cloud is too big, the temperature and pressure in the middle become so large that the cloud explodes, tearing itself apart. If the cloud is too small, it merely collapses and radiates away the heat of compression over a few million years. But clouds that are in the right range—from those having a few times as much mass as Jupiter to those having a mass about 100 times as large as that of our Sun—do something quite spectacular: they begin converting light elements to heavier elements, releasing enormous amounts of energy. The amount of energy generated is so large that these objects—stars, as we call them—can be seen a million billion miles away. And if a rock the size of the Earth is placed about 100 million miles away, the heat from the star is sufficient to warm the surface enough to sustain life.

The basic energy-release process in stars is often called nuclear burning, in analogy with the chemical process of oxidation, which we call burning. In ordinary burning, atoms of oxygen come together with atoms of whatever is burning—say carbon—to form a molecule like carbon dioxide. When the atoms combine, the electrons that were orbiting the separate atoms become shared between the two types of atoms, and the result can turn out to be a configuration that is more stable than the original separated state. "Stable" means that some energy has been released when the two came together, and this release of energy produces the heating that we associate with the word "burning."

Such rearrangements of electrons among atoms are chemical reactions, and they proceed at energies comparable to those needed to move electrons from one atom to another, namely a few volts per atom. (This is why batteries, which work via chemical reactions, produce a few volts.) Nuclear "chemistry," in comparison, involves reactions that

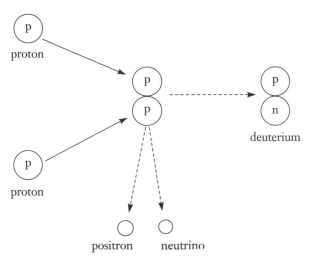

FIGURE 2.2   The first step in the proton-proton chain that powers the Sun.
Two protons come together to form deuterium, via fusion, followed by beta
decay in which a positron and a neutrino are emitted by a proton, changing it
into a neutron.

change the state of the nucleus, and this typically requires *millions* of
volts. Figure 2.2 shows such a reaction, the combination of two protons
to form deuterium. The reason that nuclear processes involve such
huge energies is that the force holding the nucleus together is much
stronger than the electromagnetic force that holds the electrons in
place. The amount of energy involved when a force acts depends on the
strength of the force, and the millions of volts needed for changes in
the nucleus is a reflection of the strength of the nuclear force.

Fortunately, in contrast to forces like gravity and the electrical attraction between charged particles, nuclear forces are short range,
which means that two bits of matter cannot interact with each other via
this force until they get very close together. Gravity is a weak force, but
its effect extends indefinitely far away, so that the Sun can hold the
Earth in orbit around it, and the mass of stars at the center of our galaxy many thousands of trillions of miles away can hold the Sun in orbit
around it. Electrical forces are also long range and they are far stronger
than gravity, but we do not normally notice them because matter is, on
the whole, electrically neutral, so that the attractions and repulsions

nearly balance and the strength of this force does not show itself. If matter were not so nearly neutral, electrical forces instead of gravitational forces would dominate our world.

How close must particles get before they interact via the strong nuclear force? The nuclear force decreases in strength very quickly at a distance of a few trillionths of an inch. This is a small fraction of the size of an atom—very short-range indeed by human standards. In order for nuclear reactions to occur, the particles involved must be brought very, very close together.

The problem is that the basic particle involved, the proton, is positively charged and positive charges repel each other. The force of repulsion between charges becomes larger and larger as they are brought closer and closer together. The amount of energy needed to overcome this repulsion and bring the protons very close together turns out to be quite large, and only if the protons are thrown together quite violently can they overcome the electrical repulsion. This was a major technical obstacle in trying to produce the hydrogen bomb, which works by fusion, and it needed the explosive force of a fission bomb to crush the atoms of fuel close enough together to initiate fusion reactions; ordinary chemical explosives were not powerful enough.

The other way to bring the nucleons close together is by high temperature, since what we mean by "high temperature" is that the particles have very high average speed. When they collide at high speed, they come close together before the electrical repulsion drives them apart again. At high enough temperatures—several million degrees— the protons get close enough together for the nuclear force to act, and fusion can take place.

Actually, however, the nuclear particles inside the Sun do not get close enough together to interact. Yet they do interact. The resolution of this apparent paradox goes to the heart of the strange and counter-intuitive quantum theory.

Aside from jokes about misplacing one's glasses, ordinary objects in our world are either there or not there. If you place a cup on a saucer, you don't have to check whether it suddenly moves to the other side of the table, or whether the saucer decides to change places and appear on top of the cup. Macroscopic objects are well localized, and we are used to things being that way. This is one of the reasons that we have such a hard time picturing the world at the atomic and subatomic levels: our sensory apparatus did not develop with the need to deal with the (to us) bizarre behavior of matter on very small spatial scales.

What is the behavior that we would find so odd? Matter, it turns out, does not exist "at a place," but rather exists as a set of probabilities of being in a range of places. This is the famous wavelike nature of particles, and the usual analogy is made with a wave on a string, by pointing out that the entire wave has a certain size and it makes no sense to localize it to any smaller length than the distance between one wave and the next. This picture is not quite satisfactory, because we can look at a vibrating string and we can see the entire piece of string waving. We can point to each piece of it as being well localized: we can see that the string or wave has parts and can identify each of the parts as it moves back and forth or up and down. In quantum physics, this is not possible. A small bit of matter really exists only as a cloud of probability until it interacts with another bit of matter (or radiation) at a particular place. Until the interaction occurs, the particle could be anywhere, although with decreasing probability as you search regions of space far from the most likely locations.

As with many of the other predictions of quantum mechanics, this one is hard to accept (Niels Bohr, one of the key figures in the development of quantum mechanics, remarked that if you are not appalled by the theory then you haven't understood it). What it means for our nuclear burning problem is that, when two protons are thrown at each other, there is a small probability that they are actually in the same place, even if they cannot come all the way together under a classical (nonquantum) calculation. This effect was first called barrier penetration—although these days it is often called tunneling—because the problem is formulated mathematically as one in which there is a barrier holding the two particles apart. Like a convict attempting to escape prison, the particle can tunnel under the barrier (metaphorically) and appear on the other side. It is even possible to calculate quite accurately the probability of tunneling, and such calculations were indeed carried out in the late 1920s and early 1930s.

The probability of a particle appearing on the other side of a barrier is very strongly dependent on the difference between the energy of the particle and the energy needed to penetrate the barrier. The energy of the particle is directly tied to the temperature in the center of the star, since the average energy per particle is what we mean by temperature, so the penetration probability and therefore the rate at which the fusion reactions take place are very strongly dependent on the temperature. This is why small stars, with moderate temperatures in their cores, are so faint and live many billions of years: their fusion rate is relatively low. Large stars, with high central temperatures, are exceedingly bright and burn up their fuel in only a few million years.

The fusion of light particles into larger clumps of so-called heavy nuclei is responsible for the formation of the elements. But a complete understanding of the processes involved in forming the elements took several decades to achieve, from George Gamow's work in 1928 until the late 1950s. It seems that there are three major nuclear cooking-pots in which element-building takes place. Two of the three directly involve stars:

1. The first 3 minutes. During the first few minutes of the Big Bang, the temperature and density of the expanding fireball were high enough for the fusion of protons and neutrons into heavier elements to occur, but cool enough (less than a billion degrees) that the heavier nuclei were not torn apart as soon as they formed. This happened at the age of about 1 minute; by the age of 3 minutes, the universe had expanded and cooled enough so that the formation of heavy elements stopped. The basic process was worked out by Ralph Alpher in 1948, in collaboration with Gamow. (The name of the nuclear physicist

Hans Bethe was added by Gamow for aesthetic reasons, so that the list of authors sounded like the first three Greek letters: alpha, beta, gamma. Gamow liked to play jokes.) Only the lightest elements formed this way: hydrogen (1 proton), deuterium (1 proton and 1 neutron), some tritium (1 p and 2 n), a large amount of helium (2 p and 2 n), a small amount of light helium (2 p and 1 n), and very small amounts of lithium (3 p) and beryllium (4 p), formed by the combination of deuterium and tritium. The problem with getting elements beyond helium in any great numbers is that the next nucleus, having a mass of 5, is unstable and doesn't live long enough for a sixth nucleon to join in, except in rare multiparticle interactions. So the heavier elements must have been formed after the Big Bang.

2. Stellar interiors. Largely through the work of Hans Bethe in 1938, the fusion of light nuclei into heavier ones in the cores of stars was understood. With this development, the energy source of stars was explained, and a first step was taken in explaining the formation of some heavy elements out of lighter ones. The basic processes, taking protons into alpha particles (helium nuclei) and other steps that made elements into carbon, were understood. Carbon, with 12 nucleons (protons and neutrons), can be made by fusing 3 helium nuclei, for example, since each helium nucleus contains 4 nucleons.

3. Novas and supernovas. In 1957, a massive survey of many possible fusion processes was carried out by E. Margaret Burbidge, Geoffrey Burbidge, William A. Fowler, and Fred Hoyle (and independently by Alastair G. W. Cameron). This included processes not occurring in ordinary stellar interiors, such as the rapid (r) and slow (s) neutron fusion reactions, which were later understood to occur in nova and supernova explosions. In this way, the heavy elements, through uranium and the even heavier trans-uranic nuclei, could be produced in these massive stellar explosions. Many of these heavy nuclei decay very rapidly, or at least rapidly compared to the age of the universe, and are not normally found in nature now.

So it is true that we are stardust: the elements that make up our bodies, and nearly everything we see around us, were manufactured inside stars.

THE EVOLUTION OF SOLAR ACTIVITY

Although you can't (and shouldn't) usually look at the Sun directly, sometimes at sunset it is dimmed just enough by distant haze that you can safely see what is on it. Occasionally, a small dark spot can be seen, even with the naked eye.

These dark regions, called sunspots, were studied in 1610 by Galileo, who turned the newly invented telescope on objects in the sky (Figure 2.3). About the same time, Christopher Scheiner, a Jesuit astronomer, saw the sunspots, and a lively debate over their origin and significance ensued. We now know that the sunspots are relatively cool regions on the Sun, over 1,000 kelvins cooler than the surrounding solar surface—still hot, by terrestrial standards, of course. As a consequence, the sunspot gas is less bright than the surrounding gas, appearing dark in comparison. The number of spots on the Sun increases and decreases from year to year, with an average of about 11 years from one maximum to the next; this regular variation is called the sunspot cycle.

Early in the twentieth century, George Ellery Hale examined the spectra of sunspots. He discovered that certain spectral lines were split into three parts: instead of seeing just a single sharp line in the spectrum, you see the expected line plus an additional line on either side of it. Such splitting had been discovered earlier in a laboratory setting by the Dutch physicist Pieter Zeeman. It is due to the presence of a strong magnetic field, which alters the energy levels in some atoms, so that the atomic transitions which absorb and emit light are slightly changed in energy. The magnetic field shifts the line both higher and lower in energy, so that the splitting can show up on both sides of the original line. Stronger magnetic fields result in greater splitting of the levels. In the Sun, Hale attributed the splitting in sunspots to a magnetic field of

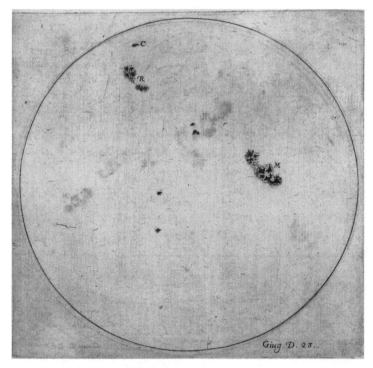

FIGURE 2.3 Drawing of sunspots made by Galileo in 1612. His sequences of such drawings showed that the Sun was rotating and that these blemishes were on the surface of the Sun, which was therefore not "perfect."

3,000 gauss, about 3,000 times higher than the 1-gauss average magnetic field on the Earth (Figure 2.4).

The strong magnetic fields in sunspots limit the conduction of heat flowing up from the interior of the Sun, diverting the upflow from the interior and keeping the spots relatively cool. We can even detect, from their spectra, several molecules in sunspots, while the normal solar surface is so hot that molecules are broken apart.

About a hundred years ago, E. Walter Maunder in England noticed that at the start of a sunspot cycle, sunspots initially emerge at high latitudes from the solar equator. Then, as the years go by, new spots form closer and closer to the equator. When they get within about 15 degrees of the equator, a new set of sunspots starts forming again at

FIGURE 2.4   Measuring the magnetic field in a sunspot via the Zeeman effect.
The strong spot field splits the line into three components.

higher latitudes. He plotted the effect in what has become known, for
obvious reasons, as the butterfly diagram (Figure 2.5). You can see that
as the 11-year cycle continues (from left to right), sunspots first appear
at a high latitude and then appear at lower latitudes. The appearance of
new spots at high latitude marks the beginning of a new cycle.

Scientists are unable at present to predict how strong a cycle will be
in terms of sunspot number. For example, a set of papers in the late
1990s tried to predict how high the peak of the cycle would be in the
year 2000. Of the set, one paper predicted that it would be higher than
average, another predicted that it would be lower than average, and a
third predicted that it would be average. It has also been suggested that
there are other cycles with longer periods than 11 years that are superimposed on the 11-year cycle. The evidence for these longer cycles is
still marginal, and unless we can find a way to extend the historical sunspot record back before 1610 we will just have to wait longer to obtain
higher statistical certainty for such cycles.

In the meantime, the 11-year sunspot cycle is very obvious. Later in

FIGURE 2.5 The butterfly diagram, the location and number of sunspots appearing on the Sun plotted against time for the most recent solar cycles, with 1977–2000 shown.

this book, we will see that the sunspot cycle is only a symptom of a more general solar activity cycle. What drives the sunspot cycle is a cycle in the intensity of the Sun's magnetic field strength. These magnetic fields, generated inside the Sun, create a large assortment of dynamic effects that are collectively called the solar activity cycle. The sunspot cycle is the most easily observed manifestation of the activity cycle.

Maunder also noticed something puzzling about the historical sunspot record: there seemed to have been very few spots on the Sun in the decades after Galileo's discovery. Sunspots were generally absent for about 70 years, coinciding with the reign of France's "Sun King," Louis XIV, and returning in large numbers after his death in 1715. Maunder's work was forgotten for decades and then rediscovered by John Eddy of the High Altitude Observatory in the 1970s. This "Maunder minimum" may indicate that the sunspot cycle turned off for a time in the 1600s, which would imply that the sunspot cycle is superficial rather

FIGURE 2.6   Early-seventeenth-century painting by Hendrick Avercamp (1585–1634) showing ice skating on a canal in Holland during a period that many think was an early part of the Maunder minimum. The canals very rarely freeze over in modern times.

than fundamental to the Sun. Long-period cycles like this may turn out to be important for understanding the linkage between the Sun and the weather on Earth. In particular, the Maunder minimum corresponded to a relatively cool time in Europe sometimes known as the Little Ice Age (Figure 2.6). The southwestern United States was in drought at the same time.

Were there really fewer sunspots during 1645–1715, as Maunder and Eddy suggest? Or were people just not watching closely enough? The fact that individual papers appeared reporting on sunspots indicates that they did indeed attract attention, and would have been written about. Contemporary reports state that observers had been searching in vain for years to see even a single spot on the Sun. Further, other indicators of the Sun's magnetic field also were weak or absent during much of that time. For example, records show that there were few auroras and that the corona of the Sun was not strong when seen at eclipses during that interval.

In the 1980s, Sallie Baliunas built on work started by Olin Wilson and others at the Mt. Wilson Observatory, establishing that several other stars also have solar activity cycles. These observers followed a certain spectral feature that is typical of solar activity yet is visible from the ground, and noticed that the strength of this feature varied in nearby solar-type stars with periods ranging from a few years to 15 years or so. Some stars may have longer periods, but more years of observation are needed to be sure of that. They also found some solar-type stars that appeared to be in a Maunder minimum phase of very low activity. Enough observations of this type could establish how much of the time the Sun is in a protracted minimum of activity, by determining what fraction of the solar-like stars are in such a state at any given time. This work is a good example of the solar-stellar connection, the use of solar and stellar data to complement each other: our knowledge of the Sun helps us to interpret the stellar data, and the stellar observations help us test theories that are developed about the Sun.

THE SUN TODAY

The Sun is in the prime of its life, about halfway through its 10-billion-year history. We expect slow changes in the Sun during this time, especially as the hydrogen fuel in its core is gradually converted to helium and heavier elements. But we also know that there is the very rapid 11-year sunspot cycle, and perhaps other cycles as well.

There are indications that such short-term (compared to the 10-billion-year lifetime) variations in the Sun show up as changes in the Earth's climate or weather patterns. So we may naturally wonder whether they show up in the total energy output from the Sun. In Chapter 7 of this book, we will see how the American scientist Charles Greeley Abbot tried for decades to ascertain if the sun shines steadily.

The accuracy necessary to detect variations in the Sun's energy output was beyond his capabilities, but with the launch of space satellites, we began to be able to make the appropriate measurement.

The solar constant is the amount of energy that reaches each square meter at the radius of the Earth's orbit. It is taken to be the amount that reaches the top of our atmosphere, so that we can measure it without regard for the vagaries of atmospheric effects. Starting with the Solar Maximum Mission (SMM) in the 1980s, a series of satellites has been launched with instruments able to take in and measure all of the solar energy, unfiltered by terrestrial atmosphere. The instrument used—essentially a cavity that lets radiation in and then measures the resulting temperature of the cavity—improved the ground-based accuracy of 1.5 per cent to the current 0.1 per cent. The result is about 1,368 watts per square meter (Figure 2.7).

But the spacecraft also found deviations of about 0.2 per cent, often coinciding with the passage of a large sunspot, or spot group, across the Sun. Only with that measurement did it become known that sunspots actually diminish the solar radiation, instead of just redistributing it to other parts of the surface. Frightening for a time was the discovery that the solar constant was diminishing at a rate that would be serious for us on Earth if it continued for only a few decades, very short on an astronomical time-scale. Fortunately, when measurements were made over enough years, we saw that the overall change in the solar constant was linked to the solar activity cycle. When the sunspot number started rising, the solar constant started rising, until it reached the value seen at the previous peak of the cycle. Several spacecraft, notably the Solar and Heliospheric Observatory (SOHO) with its VIRGO experiment, continue to monitor the solar constant—which isn't really a constant at all, and whose changes may be linked to changes in the Earth's weather patterns, as we'll discuss in Chapter 7.

The sunspot number continues to be counted. It is tabulated by the International Sunspot Data Center in Belgium, which averages the reports of dozens of astronomical sites. Another tabulation is carried out by the American Association of Variable Star Observers in the

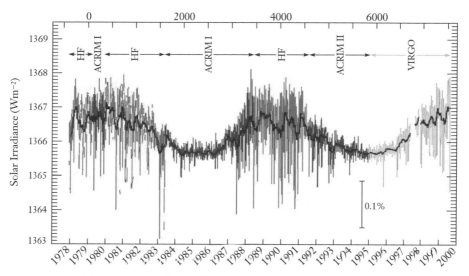

FIGURE 2.7 Measurements of the solar constant over nearly two sunspot cycles from four different satellites. VIRGO on SOHO is also making measurements.

United States, which averages reports made by many amateur astronomers. These counts are available on the Internet each month at http://www.oma.be/KSB-ORB/SIDC/index.html (Figure 2.8), as is the butterfly diagram. The "sunspot number" has a historical significance, since it has been kept for about 150 years and can therefore be used to extend the range of certain studies back in time. But it isn't really the number of sunspots, which after all come in a wide range of sizes. The mere number of spots may not be as significant as, say, the total area of the solar surface covered by spots.

But measuring the area covered by spots is very difficult and tedious, so that a compromise measurement is used. The "sunspot number" is defined as ten times the number of groups of spots on the Sun plus the number of individual spots, all multiplied by a "personal constant." The personal constant takes account of each individual observer's tendency to be biased in a consistent way, such as how he or she decides whether there are one or two spots in an ambiguous region, and of the fact that such tendencies vary from one individual to another.

Let us consider that there is one spot on the side of the Sun facing us.

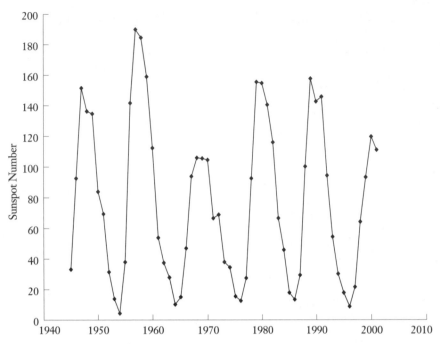

FIGURE 2.8   The monthly averaged sunspot number for the last five solar cycles, from the Sunspot Index Data Center at the Royal Observatory of Belgium.

(The sunspot number does not take account at all of the fact that we are seeing only half the Sun at a time; spots on the other side simply are not part of the count.) If there is 1 sunspot present, then the sunspot number is 1 group + 1 spot and the sunspot number is 10 + 1 = 11. If there are 2 sunspots present, it matters how they are distributed. If they are together, then there is 1 group + 2 spots and the sunspot number is 10 + 2 = 12. If they are separate, though, then there are 2 groups + 2 spots and the sunspot number is 20 + 2 = 22. The sunspot number, as you saw in the figure, can be as low as 0 and, averaged over a month, can exceed 180. For an individual day, the sunspot number can reach 250 or greater.

Some graphs plot the average daily area of the Sun covered with sunspots, calculated as a percentage of the entire solar disk facing us. A typical sunspot covers about 0.02 per cent of the disk, so that at most half a per cent of the Sun might be covered. More typically, only about a tenth of a per cent is occupied by spots. If our Sun were at the distance of other stars, we would not detect this activity, whereas we do actually see noticeable spot-related brightness changes on several nearby stars. Thus these stars must have much more coverage by spots, and our Sun seems to be relatively low in activity compared to its neighbors.

THE FUTURE

The Sun is reasonably steady now, but starting in about 3 billion years, changes in its core will be reflected in a change at the surface. The outer atmosphere of the Sun will begin to swell so much that our own atmosphere will heat up, forcing us to leave the Earth. We on Earth would eventually be incinerated if we stayed. Fortunately, we have billions of years to arrange to go elsewhere.

Eventually, the Sun's outer layers will brighten and redden faster and faster. These outer layers will envelop even the current orbits of Mercury and Venus, reaching approximately the current orbit of the Earth. The Sun will then be a type of star known as a red giant. We know of many such stars in the sky. It will be red because the outer layers will be relatively cool, a bit more than half the temperature they are today. But even though the solar surface will be cool, it will be so close to Earth that oceans will boil and humans could not survive. The solar wind, an outflow of particles, will be so strong that the Sun will be losing mass rapidly, which will make the planets recede a bit, but not enough to save us.

While the outer layers of the Sun are enlarging, the core of the Sun will be collapsing inward, which will raise the central temperature and pressure. By about 5 billion years from now, the core will have risen from its current 15 million degrees to about 100 million degrees. At that point, helium atoms will begin to fuse together three at a time, first making beryllium, then carbon. This is known as the triple alpha process and it produces what is known as the helium flash, because the process commences very rapidly and exhausts itself in such a short time interval, as short as a few seconds for some stars. The energy from the helium flash will expand the Sun's core. Because it then gives off less energy for a while after the flash ends, the outer layers will shrink again. At this time, helium will fuse into carbon in the core in a steady fashion. Outside the resulting core of carbon, shells of fusing helium and, farther out, fusing hydrogen, will exist. This stage will last a few hundred million years.

The shells of fusing hydrogen and helium will burn unstably, causing pulsations that continually increase in strength, and the stellar wind will increase by a billion times, carrying out large quantities of material into interstellar space. The material carried out, whether by pulsing or by the unstable solar wind, will drift away from the Sun. As the outflowing wind increases in strength and speed, new material will catch up with the old, piling up the material in shells around the Sun.

We see hundreds of such stars in the sky. They are known as planetary nebulas, though the word "planetary" means only that a hundred or so years ago they appeared in the telescopes of their time as fuzzy disks like the planets Uranus and Neptune. Actually, they are the clouds of gas around dying stars that were once like our Sun. Images taken by the Hubble Space Telescope show the many different forms that planetary nebulas can take, depending on the details of the irregular pulsing of the gas and on whether or not there are directional jets of gas present, aligned to the stars' original spin axes (Plate I). The Ring Nebula in the constellation Lyra is a planetary nebula familiar to amateur astronomers, since it is visible in even small telescopes.

The planetary nebula stage of the Sun's future will last only about 1,000 years, a snap of a finger in the Sun's 10-billion-year overall lifetime. After that time, the gas surrounding the star will expand and cool so much that it will become invisible. It will carry the carbon and other elements out into space, providing elements necessary for life forms such as ourselves.

The core of this remnant Sun will cool and contract, until the electrons inside it cannot be pushed together any more by the gravity of the remnant's mass. Our Sun will then be an Earth-sized star, of the type known as a white dwarf, and will end its life in this way, slowly cooling until the end of time.

What We See

The Sun is basically a hot ball of gas, powered by nuclear reactions at its core and eventually radiating that power out into space at its outer surface. The temperature at the core is millions of degrees, so that the "light" produced there is mainly X-rays. But this light must travel through an enormous amount of matter to get to the surface of the Sun, being scattered, absorbed and reemitted so many times that an astounding 100,000 years are needed for the energy generated in the core to get to the surface.

This energy is spread over a larger and larger area as it moves outward. The average temperature of the radiation drops as it moves out, but in such a way that the total—the local intensity multiplied by the area of the larger and larger surface—remains a constant. By the time the energy reaches the visible surface of the Sun, the temperature has dropped to about 5,800 K (10,000°F). This puts the radiation emitted into the visible part of the spectrum—our eyes almost certainly having adapted to the available wavelengths, thereby making them the "visible" part.

THE PHOTOSPHERE

Every day that is not cloudy, we see a bright disk of light in the sky. Visually, the Sun seems to have a surface, and the light we see comes from that surface. "Photos" is the Greek word for "light," and "sphere" accounts for the spherical shape of the solar surface, so astronomers use the term "photosphere" to mean that visible surface of light.

But if the Sun is made of gas through and through, how can there be anything that we call the Sun's surface? And why does the Sun seem to have a sharp edge? To understand the answers, we need to know some of the properties of gases. Most gases appear transparent, but we perceive this seeming transparency only because we are looking through gases for small distances. Imagine a smoke-filled room, for example (a location that does not exist in quantity any more): we can usually see our hands at the ends of our arms, but may not be able to see the far wall. The smoke is opaque enough that we cannot see through 20 feet of it.

We measure how opaque something is with a quantity named "opacity," the opposite of transparency. Something that has opacity 0 is completely transparent. We can see only murkily through something that has opacity 1. And by the time the opacity is 5 or more, we effectively can't see through it at all. (Technically, when light goes through a gas of opacity 1, its intensity is diminished by a factor of e, a natural constant equal to 2.71828 . . . Opacity 2 is a factor of e^2 or roughly 7; and so on. Opacity 5 is a factor of 148, so less than 1 per cent of the light gets through.)

So when we look at the center of the Sun's disk, we see into the gas until the opacity builds up enough so that light from a lower level effectively doesn't get out. It turns out that most of the light we receive is from a level at which the opacity is 2/3. (Two thirds is not quite as far inward as opacity 1.) We get a little light from farther out, and a little light from farther in, but it is convenient to talk of the level at which the opacity is 2/3 as the surface of the Sun. The photosphere is a few thousand miles thick, but we define the level at which the opacity is 2/3 as the base of the photosphere.

The Edge of the Sun

When you can see the disk of the Sun—perhaps dimmed by fog or haze at sunset, or when you look through a suitable, safe solar filter—its edge seems to be sharp. How sharp is it?

In the case of the edge of the Sun, it doesn't have to be actually sharp; it only has to be sharp enough to seem that way to the human eye. The pupil in your eye through which the light comes is only a few millimeters across—sometimes only a millimeter or 2 in bright light and rarely more than 8 millimeters. The sharpness of an image in an optical system is limited by its size because of diffraction, with larger optics being needed for higher resolution. Your eye can resolve only about a minute of arc, an angular measure that is about 1/30 the diameter of the Sun or Moon.

The question to ask is: what is the angle between a line of sight that hits the disk of the Sun (the photosphere) and a line of sight that seems to go past the Sun into space farther out, as shown in Figure 3.1? That angle turns out to be less than 1 minute of arc, and so isn't resolved by the human eye. Thus the edge of the Sun—the edge of the photosphere—is narrow enough that it appears sharp to us. In reality, the edge is "fuzzy" by about 100 miles, which is large by human standards, but small on an object as large as the Sun. The same effect also makes clouds in the Earth's atmosphere appear to have sharp edges.

Limb Darkening

On a photograph of the photosphere, you may also notice that the Sun does not appear uniform in brightness across its disk: the parts of the disk near the edge appear somewhat darker. Since the edge of the Sun (or moon or star) is called its limb, the phenomenon is known to astronomers as limb darkening (Figure 3.2).

You can see the effect if you use one side of a pair of binoculars to project a solar image onto a surface. (Be careful not to look up at the

FɪɢᴜʀE 3.1   The edge of the Sun (shown as a negative), and why it seems sharp to the unaided eye.

Sun through the binoculars; use the binoculars only to project the solar image onto a screen.) The image will show sunspots and, near the edge of the Sun's surface, will look darker.

We understand the phenomenon of limb darkening in terms of the idea above, that the Sun is a murky gas. When we look toward the center of the Sun, we see in until our line-of-sight penetrates enough material to make the opacity add up to about 2/3—until the gas becomes effectively opaque. But when we look near the Sun's limb, we are look-

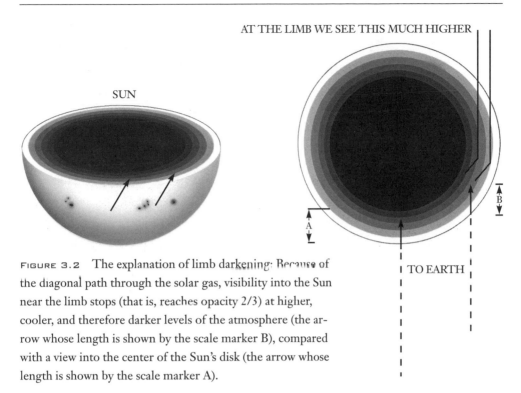

FIGURE 3.2 The explanation of limb darkening: Because of the diagonal path through the solar gas, visibility into the Sun near the limb stops (that is, reaches opacity 2/3) at higher, cooler, and therefore darker levels of the atmosphere (the arrow whose length is shown by the scale marker B), compared with a view into the center of the Sun's disk (the arrow whose length is shown by the scale marker A).

ing diagonally through the Sun's gas and have to penetrate more material to see down to a given depth below the surface. This means that the opacity adds up to about 2/3 at a higher level, farther from the Sun's center. So when we look near the limb, we are seeing photospheric gas from a higher atmospheric level than when we look near the center of the Sun's disk.

Why is the limb darker? Because the upper level of the Sun's atmosphere is cooler than the deeper level, so the light emitted from the higher levels is less intense. What we learn from studying the limb darkening in the visible part of the spectrum is that from the base of the photosphere upward for a few thousand miles, the Sun's atmosphere is getting cooler.

In other parts of the spectrum, the same effect occurs, in that near the Sun's limb we are getting radiation from higher, cooler levels. But in some parts of the spectrum, such as the millimeter-wavelength radio radiation, our instruments detect a limb *brightening* instead of a limb darkening. This is a first indication that something other than a decrease of temperature with height happens above the photosphere, and we will have much to say about this in the rest of this book.

The Temperature of the Sun's Surface

The Sun gives off what we call white light: all the colors of the rainbow. Our eyes naturally adapt to sunlight, so we see a white sheet of paper as white no matter whether it is midday, with the Sun overhead, or sunset, with the sunlight that filters through the atmosphere becoming reddish. But color film is less forgiving—photographers often find their sunset pictures looking excessively reddish, or they use a bluish filter to remove some of that annoying reddish tinge. So we know that sunlight contains red in it, and of course the blue of the sky comes from blue wavelengths that were removed from the incoming sunlight while the reddened portion came through.

Visible light, the light we can see with our eyes, ranges in wavelength from about 400 nanometers (nm) to 700 nanometers, where 1,000 nanometers is 1 micrometer, which is 1/1000 of a millimeter. In the metric system, milli- (a prefixed m) means 1/1000; micro- (whose symbol is a Greek mu) means 1 millionth; and nano- (a prefixed n) means 1 billionth; m written by itself is the symbol for a meter. We are talking, in customary American units, of about a millionth of a yard for a wavelength of light.

The distribution of the Sun's energy looks like a bell-shaped curve with a peak in the yellow-green region of the spectrum. Such a curve was shown in Fraunhofer's historical measurement of the distribution of the energy from the Sun (Figure 1.1). Intensity distribution curves of this sort for theoretically perfect hot, radiating bodies at different temperatures are known as black-body curves. Depending on the temperature, their peaks fall at different wavelengths. Black-body curves that resemble the distribution of the Sun's energy come from gases of about 5,800 kelvins (to convert to degrees Celsius, subtract 273, giving about 5,500 degrees Celsius) corresponding to about 10,000 degrees Fahrenheit. Note that the temperatures are usually given only to one or two significant digits, with zeros following; we often don't know them more accurately than that, and adding extra nonzero digits merely gives a false impression of accuracy.

A black body is a theoretical construct, an object that doesn't really exist but that would give off a standard distribution of radiation. The Sun is surely not a black body; after all, it changes in temperature from place to place, both inward and outward and from side to side through its atmosphere, whereas a black body is, by definition, at one specific temperature. But in visible light, the radiation from the photosphere is close to that of a black body. This closeness allows us to assign a temperature to the level of the solar photosphere from which most of the Sun's light comes.

THE SPECTRUM OF THE SUN'S SURFACE

The overall distribution of the continuous band of colors in the Sun's spectrum can be used to find the photosphere's temperature, as we just saw. But examination of the solar spectrum in great detail gives much more information.

The Sun's spectrum was first described in modern terms by William Wollaston in 1804, who saw dark divisions separating a few colors in the spectrum. But technology advanced soon thereafter, and in 1814 Josef Fraunhofer, in Germany, provided the spectrum that we saw in Figure 1.1. Fraunhofer noticed that the continuous rainbow of the Sun's radiation is crossed by dark lines in which the strength of the radiation is greatly diminished. These regions are known as absorption lines or, usually, Fraunhofer lines (Figure 3.3).

The spectrum is really the image of a slab of incoming sunlight spread out from side to side according to wavelength. The slab of light is produced by a slit in a metal surface on which the Sun's image is allowed to fall from a telescope. The light that passes through the slit enters a device called a spectrograph, which spreads out the spectrum ("spectro-") and records it ("-graph"). The Sun is so bright that its light can be spread out to a very great degree, while the intensity at each

FIGURE 3.3   Fraunhofer's original spectrum, shown on a German postage stamp. The spectrum is crossed by the dark absorption lines that Fraunhofer accurately mapped in the Sun's spectrum.

wavelength remains bright enough that it can be recorded. Figure 3.4 shows a solar spectrum so long that it is folded into several strips, each covering only a few nanometers of wavelength. The most detailed of such observations have been used to measure about a million Fraunhofer lines. All other stars also have absorption lines, though they are not bright enough for us to spread out their spectra as far, and not as many lines are known. Millions of distant stars, though, seem to have spectra just like that of the Sun. These stars are known as type G2 dwarfs, and the Sun is quite typical of the class.

To understand what causes the dark Fraunhofer lines, we must consider what happens when light passes through a gas. The key process is excitation or ionization of the atoms of the gas by the light coming through it from below. When the overlying gas is cooler than the source of the light (which is the underlying gas, which emits a continuous broad range of wavelengths), the energy from the light can be absorbed by the atoms in the cooler gas. These atoms take up the energy at only a few particular wavelengths for each type of atom.

Energy is neither created nor destroyed in the atoms, but the energy removed from going straight ahead is radiated by the atom in all directions. Thus less of the energy, at a few specific wavelengths out of the broad continuum of wavelengths, is left to go straight ahead. When we look back through the cooler gas at the hotter source, we see a set of dark lines subtracted out of the continuum light. The set of wave-

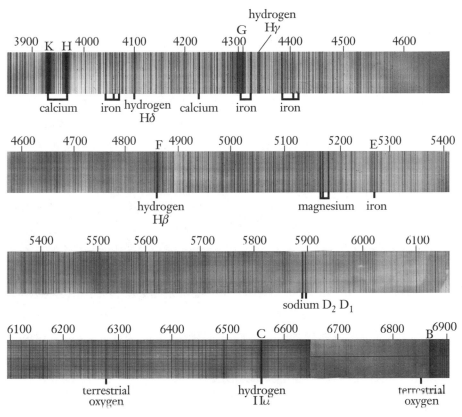

FIGURE 3.4 A solar spectrum with some prominent spectral lines identified. The wavelength is in angstroms; 3,900 angstroms = 400 nanometers.

lengths taken up for each type of atom at a given temperature acts like a fingerprint in allowing scientists to identify what atoms are present and how hot they are.

In the Sun, the fact that we see Fraunhofer lines tells us that we are looking through a cooler gas at a hotter one. This conclusion is the same one we reached from the study of limb darkening. Indeed, the temperature that we measure from the Fraunhofer lines is roughly the same as that we get from the continuous spectrum. Differences in detail arise because the Sun is neither homogeneous nor a true black body.

Some of the most prominent Fraunhofer lines that we see in the solar spectrum are from hydrogen. It might seem that the hydrogen lines should be especially prominent since 90 per cent of the gas in the Sun (and in the Universe in general) is hydrogen. But it turns out that the strongest hydrogen Fraunhofer lines are at ultraviolet wavelengths too short for us to see in visible light. The Fraunhofer lines we see in the visible spectrum are secondary lines, still strong enough to be prominent but not strong enough to be very prominent. Indeed, when Cecilia Payne (later Payne-Gaposchkin) suggested, at Harvard in 1929, that hydrogen was the dominant gas in the Sun and stars, her conclusion was quickly rejected by the then most prominent astronomers of the time. It took years before the world in general came around to her conclusion. She became used to being a pioneer—she was, much later, the first female to be promoted to professor in the Faculty of Arts and Sciences at Harvard University.

What is the hydrogen spectrum like in the visible wavelengths? Since hydrogen is the simplest element, having just one electron, its spectrum is also fairly simple. It is a series of lines, starting with a bright red line, then a less bright blue line, and then other lines still less bright in a sequence converging at shorter and shorter wavelengths through the blue and violet (Figure 3.5). These lines were analyzed in the nineteenth century by the Swiss schoolteacher Johannes Balmer, and are still known as Balmer lines.

The strongest Balmer line, the red line, is known as H-alpha. The other lines are lettered, continuing in the Greek alphabet: H-beta, H-gamma, and so on. Whenever you spot that sequence of converging absorption lines, you know that hydrogen is present. By analyzing the strengths of the lines, you can often tell how much hydrogen is present and how hot the gas is. Balmer realized that each state of the hydrogen atom at a certain temperature corresponded to a certain amount of energy, and that the series of lines we now know as Balmer lines corresponded to changes in the state of the atom from one well-defined energy to another. We now know that these changes come from electrons jumping from one energy level in the atom to another.

The strongest Fraunhofer lines in the visible spectrum are a pair of purple absorption lines, just into the ultraviolet at the limit of our eyes'

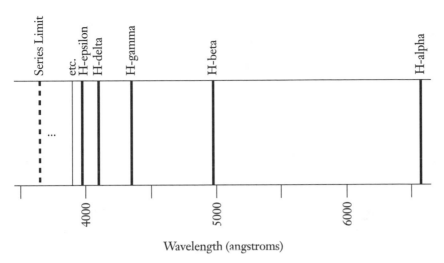

Wavelength (angstroms)

FIGURE 3.5 The spectrum of hydrogen at visible wavelengths, showing the first six lines in the series named after Balmer.

visibility. Fraunhofer had labeled the strongest lines in his spectrum A, B, C (which turns out to be H-alpha), and so on, and one of these strongest lines is Fraunhofer's H-line. Fraunhofer had used the letter I to indicate the end of his spectrum, but someone unknown later in the nineteenth century had associated the letter K with the second of the two strong ultraviolet lines. So these "H and K lines" are the pair of very strong absorption features in the violet spectrum of the Sun. They turn out to be from once-ionized calcium, calcium that has lost one electron. They are the fundamental lines that are absorbed by a calcium atom, which makes them very strong, given that calcium is a reasonably abundant element on the Sun. They are less strong than the fundamental hydrogen lines that we can't see in the ultraviolet spectrum, though stronger than the Balmer series lines of hydrogen that we see in the visible part of the spectrum.

Most of the lines that appear in the Sun's spectrum are from iron and from similar elements. Iron has so many electrons (26 in the un-ionized state) that the spectrum resulting from the various jumping around of electrons, so simple for hydrogen, is very complex in iron. None of the individual iron lines is exceptionally strong, though. Still, some have

FIGURE 3.6 Surface magnetic fields on the Sun at the solar maximum in 1979 (left) and at the minimum in 1986 (right).

special uses in astronomy, particularly those that are sensitive to magnetic fields (Figure 3.6), which are used to map the strength of the magnetic field.

Historically, people spoke of the Sun's "reversing layer": hot gas at the base of the photosphere and cooler gas above it causing the Fraunhofer lines at higher levels. We now know that the situation is not quite so simple. The Sun's continuous radiation and the formation of the Fraunhofer lines are mixed through the photospheric layers, with both forming gradually with height and with many details depending on exactly which line you are studying. The overall temperature of the photosphere seems to range from 6,400 kelvins (11,000°F) at the lowest levels to 4,400 kelvins or so (7,500°F) at the upper heights.

The Composition of the Photosphere

From the Fraunhofer spectrum, astronomers can analyze how much of each element is present. They have found almost all the known chemical elements there. The quantity of an element is known as its abundance, and most abundances astronomically are compared with that of hydrogen or with that of carbon. Normally these abundances are given by number, and not by mass. For example, if we have 9 atoms of hydrogen and 1 of helium, hydrogen is 90 per cent of the elements by number. But the hydrogen atomic mass is 1 and the helium atomic mass is 4,

Table 3.1   The most abundant elements in the solar photosphere.

For each: 1,000,000	atoms of hydrogen
There are: 98,000	atoms of helium
850	atoms of oxygen
400	atoms of carbon
120	atoms of neon
100	atoms of nitrogen
47	atoms of iron
38	atoms of magnesium
35	atoms of silicon
16	atoms of sulfur
4	atoms of argon
3	atoms of aluminum
2	atoms of calcium
2	atoms of sodium
2	atoms of nickel

so the total atomic mass of the atoms is $9 + 4 = 13$. The abundance of hydrogen by mass, if we choose to compute it that way, is thus 9/13, or about 70 per cent. (This way of measuring abundances also is used in situations closer to home. For instance, the alcohol content of wine is usually given by volume, rather than by weight. Since alcohol is lighter than water, the amount would seem lower if it were stated by weight.)

The actual calculations of the abundances depend not only on the solar measurements but also on properties of the elements themselves measured in laboratories on Earth. These latter studies are known as "laboratory astrophysics" and are fundamental to understanding the Sun. The laboratory data are continually refined, most recently providing good values for the element iron and thus an accurate value for the abundance of iron in the Sun.

The Sun, of course, is making elements deep inside its core even as we speak. These elements are not fully mixed throughout the Sun. Thus the elemental abundances we measure for the photosphere are not those for the Sun as a whole, though the general trends are the same.

Table 3.1 shows the photospheric abundances of the most common elements. Note that hydrogen and helium together make up more than 99 per cent of the atoms in the Sun: everything else is a trace element.

Indeed, astronomers refer to everything heavier than helium as "metals," simplifying some calculations so that only terms for hydrogen, helium, and metals are used. Note that this use of the word "metal" is different from the electrical-conducting and heat-conducting definition in normal usage. Helium cannot be seen in the solar photosphere, so the abundance in Table 3.1 for helium comes from measurements of the outflow of gas from the Sun known as the solar wind, in which we can detect helium directly. Similarly, the neon measurements are from other astronomical sources.

MOTIONS ON THE SUN'S SURFACE

Colors are so vaguely defined that you can't really tell when they change slightly. But Fraunhofer lines are so clearly and sharply defined that you can measure small changes in their wavelengths. These changes are caused almost entirely by the motion of the gas that is emitting the lines.

The change in wavelength caused by motion is known as the Doppler effect, after the nineteenth-century scientist Christian Doppler. It is the same effect you hear when a train goes by you, and the pitch of its whistle seems to go down. When the train is approaching you, the pulsations of air coming from the whistle pile up more frequently, and the wavelengths are scrunched up, since subsequent wave peaks are emitted when the train is closer to you than the earlier wave peaks: the frequency is higher and the wavelength is shorter. When the train is receding from you, its wavelengths are stretched, and the frequency goes down, for the opposite reason. So as the train passes, you hear the frequency go from higher to lower. That change in frequency corresponds to a change in wavelength, which we commonly call a Doppler shift.

The corresponding measurement for light is the Doppler shift that moves Fraunhofer lines to slightly longer or shorter wavelengths. Since blue is at the short end of the visible spectrum, changes toward shorter wavelengths—corresponding to motions toward you—are known as blueshifts. Since red is at the long end of the visible spectrum, changes

FIGURE 3.7   A spectrum showing left and right displacements (to shorter and longer wavelengths) due to motions of granules on the solar surface.

toward longer wavelengths—motion away from you—are known as redshifts. (Even when things are shifted through and past the red to longer wavelengths, they are still known as redshifts because of the term's historical roots.)

Remember that the Sun is so bright that we can spread its spectrum out to a tremendous degree while still leaving enough brightness to be recorded in our measuring instruments. When you look at a solar Fraunhofer line in great detail and at high magnification, you can see that some bits of it—corresponding to different nearby locations on the surface of the Sun—are shifted to longer wavelengths and others to shorter wavelengths. Thus we speak of "wiggly lines"—a technical term in this context (Figure 3.7). These wiggles are really just Doppler shifts from bits of gas on the Sun coming toward us (providing blue-shifts) or going away from us (providing redshifts). We measure velocities of about 2 kilometers per second for the wiggly lines. This is a low velocity for astronomy, though it is still 7,200 kilometers per hour, about 10 times the speed of a passenger jet. These wiggly lines are a clue that there is some type of organized motion in the photosphere, with masses of hot material moving both toward and away from us.

GRANULATION

If you look in detail at the solar photosphere on a day when the air above is steady (we say that "the seeing is good," using the term "seeing" in a technical sense to indicate the steadiness of the image), you see a roughening of the surface, known as granulation. These granules are the smallest features you can see on the Sun, given the turbulence of the air above you, except at times of unusually good seeing. This turbulence normally limits you to about 1 second of arc (1/60th of a minute of arc and therefore 1/3,600 of a degree of arc), which corresponds to about 700 kilometers (400 miles) on the Sun, about the distance from New York to Washington. Imagine the entire Eastern Seaboard of the United States blurred out to a single unresolved blob and you will get an idea of the minimum size of what we can see on the Sun. There are about 5,000,000 of these granules on the Sun at each moment.

The granules are features that are typically about 1.5 second of arc across, which means that seeing them in detail is at the limit of our capability; many of the granules are smaller. One technique used to examine them is to take a lot of short exposures very rapidly, and then select the few for which the seeing was exceptionally good. Such images show bright granules with dark lanes between them (Figure 3.8). The dark lanes may be only 1/4 arc second across. Sometimes the highest-resolution images show bright points along the dark lanes. These bright points, known as filigree, seem to be 100°C or so hotter than the surrounding surface, and may be linked with the magnetic field, showing how finely broken up the magnetic field is as it pokes through the solar surface.

The granules turn out to be blobs of gas that have just risen from lower levels of the Sun, from below the photosphere. These blobs carry energy upward, just as bubbles of water in a boiling pot carry energy upward. Thus granules are a boiling phenomenon on the Sun, with their bases perhaps only 100 miles down. After the granules reach the

FIGURE 3.8   High-resolution close-ups of the solar photosphere taken in visible light, showing the solar granulation and the filigree, which appears as bright knots between the granules.

surface, they cool and sink back down, only to be replaced by new granules after about 15 minutes. We shall see later on how granules seem to oscillate in brightness with a period of 5 minutes. That oscillation was once thought to be a minor effect, but we will see that it and other, related, oscillations are the key to understanding the Sun's interior.

To see the granules without looking through turbulent air, we must be in space, where atmospheric effects don't cause seeing problems. A mission carried aloft aboard a space shuttle, for example, provided an excellent series of images of granules and how they appear and then move slightly across the photospheric surface to merge with their neighbors. These images were used to help understand how convective

5700 km

t = 0 min   t = 3.5 min   t = 7 min

t =10.5 min   t = 14 min   t = 17.5 min

FIGURE 3.9   Granulation photo sequence, showing the buffeting endured by small magnetic elements at the solar surface.

motions evolve in the complex conditions of the Sun, where the temperature and density of the atmosphere change very rapidly with depth. One of the memorable features of these data was the observation that some of the granules were seen on the space-shuttle movies to seem to explode, with debris going out to the sides.

It is also possible from the ground to obtain granulation movies of very high quality during rare periods of exceptionally good atmospheric seeing. A sequence of such images (Figure 3.9), taken at 3.5-minute intervals at the Swedish Vacuum Solar Telescope, shows small "bright points" in the lanes between granules. These points correspond to strong concentrations of magnetic fields, and the sequence shows that they are buffeted and shredded by the convective motions. This type of rearrangement of the magnetic fields may be connected to the heating of the overlying corona.

An image of the Sun, even one merely projected onto a piece of paper, also reveals slightly bright regions, irregular in shape, which are best seen in the outer parts of the disk, where they are superimposed on the limb-darkened region just inside the edge of the disk. These regions are known as faculae. Each facula corresponds to a region on the Sun where the magnetic field is slightly higher than surrounding regions. Many terms like facula come from the early years of the twentieth century when George Ellery Hale and his colleagues were examining the Sun in detail for the first time.

We will not detail all the individual names for phenomena on the Sun, but faculae are historically important, for a reason unrelated to solar activity per se. In the 1960s, careful measurements seemed to indicate that the Sun was not quite round, with its equator bulging about 35 unaccounted-for kilometers. A team of scientists incorrectly deduced that the small extra brightness they found near the Sun's equator implied that the equator was bulging by a few kilometers. They inferred from those extra kilometers that the inside of the Sun was rotating much faster than the outer layers that we can see. This extra rotation would, in turn, have an effect on the orbit of the planet Mercury. But we don't see such an effect on Mercury! Thus the scientists concluded that Einstein's general theory of relativity, which otherwise explains Mercury's orbit nicely, had to be modified. They then had their own alternative theory waiting. It turned out, though, that their measurements that the Sun was not quite round were really measurements that the Sun's edge was slightly brighter near the equator (causing it to look "fatter") than near the poles, and that extra brightness was caused by the faculae. So the study of solar faculae turned out to have major consequences for the fundamental theoretical framework that we use to understand our Universe.

The granules don't just go up and down by themselves. They have a pattern of flow across the surface of the Sun: they form within a larger-scale pattern known as supergranulation, each cell of which is about 20,000 miles (30,000 kilometers) across—about 2 and 1/2 Earth diameters in size. These supergranules last a day or two, much longer than the 15-minute lifetimes of the individual granules. They are often spoken of as polygons, since their shapes are not just round, but neither are they very irregular.

The supergranules' horizontal motion is about 1,000 feet/second (0.4 km/s). When the granules move outward from the center of a supergranule toward its edges, they carry along bits of the Sun's magnetic field, and this magnetic field piles up at the edges of the supergranules. Supergranules thus show up in magnetic maps of the Sun as an irregular pattern, like a network, of concentrated magnetic field.

The supergranules also show up in velocity maps of the Sun (Figure 3.10), which is how they were first discovered. There is a slight upwelling at the centers of supergranules, then the horizontal flow, and a downdraft at the edges, perhaps only 1/4 the velocity of the upwelling. Doppler images show the horizontal flows, bright for motion toward the Earth and dark in the away direction. At the center of the image, when we are looking straight down at motions that are mainly perpendicular to the viewing direction, there is nearly no Doppler shift and the supergranular pattern is hard to see.

The Global Oscillation Network Group (GONG) is a ground-based set of telescopes that are positioned so that, at one station or another, they are continuously monitoring the Sun. Their data for the solar surface can be broken down into several different components, some of which show the velocities and thus the supergranulation (Plate II). These components are, respectively: I, an image of the white light intensity; D, the corrected intensity, intensity variation after the overall brightness variations have been subtracted, as measured by the darkness of a chosen absorption line; V, the velocity field, measured by the Doppler shift (this shows mainly the solar rotation, which is toward us on the left and away from us on the right, but close examination also shows the supergranulation pattern); and B, the magnetic field strength, measured via the Zeeman effect.

Recently, one of the instruments onboard the Solar and Heliospheric Spacecraft (SOHO), a joint European Space Agency/National Aeronautics and Space Administration (ESA/NASA) project, has de-

FIGURE 3.10   A velocity map of the solar surface, showing the large-scale flow pattern known as supergranulation.

tected that the outflow of hot material from the Sun known as the solar wind is emitted through the lanes between convection cells, essentially through cracks in the solar surface (Plate III).

The phenomena we have just described—including the photosphere, the granules, and the faculae—are on the Sun every day. We speak of the underlying "quiet Sun" that has these phenomena.

Sunspots, on the other hand, wax and wane with an 11-year period. They and their surrounding active regions, discussed in Chapter 2, are known as the "active Sun." So solar astronomers often talk of a basic quiet Sun with an overlay of active Sun.

The quiet Sun can be well studied in the visible part of the spectrum. But the photosphere is too cool to give off x-rays, and does not give off radio waves, so does not appear in other parts of the spectrum than the visible. On the other hand, some of the other parts of the Sun are best studied in these other parts of the spectrum, as we shall see.

THE CHROMOSPHERE

During a solar eclipse, just after the Moon entirely covers the solar disk, a reddish rim becomes visible. It is a thousand times fainter than the photosphere but still a thousand times brighter than the corona that will soon appear. This reddish rim is a shell of gas around the photosphere. From the Greek words "chromos," meaning "color," and sphere, it is called the chromosphere. The chromosphere is only about 10,000 kilometers thick, less than 1 per cent of the 1,500,000-kilometer diameter of the Sun.

At the beginning of the total phase of an eclipse, when the light fades enough, the chromosphere becomes visible even to the naked eye. Previously, we described how light from a hot source going straight ahead into a cooler gas can be absorbed by certain atoms, giving rise to the Fraunhofer absorption lines. The lines are dark because the cool gas absorbs the light trying to pass through, and then reemits that energy in all directions. So when you look down from above, as we do when we look at the photosphere, you detect the missing energy by seeing dark lines. But when you look from the side, as we do when we see the chromosphere at the edge of the Sun, we see energy emitted without the bright photospheric background. Thus we see a set of bright wavelengths coming from the chromospheric gas.

These wavelengths include all the Balmer lines of hydrogen. The brightest of them is H-alpha in the red, so the chromosphere appears reddish. But the bluer wavelengths of Balmer lines are also mixed in, as are the violet wavelengths of the calcium H and K lines and several other emission lines.

The chromosphere is covered within a few seconds by the advancing edge of the Moon. After the seconds or minutes of totality, the other edge of the Moon uncovers the chromosphere at the other limb of the Sun. When you see that reddish edge just over the Moon's limb, you know that the total phase of the eclipse is about to end. (We will say more about eclipses in Chapter 5.)

The Flash Spectrum

The laws of radiation from hot gases are such that the bright chromospheric emission lines are at exactly the same wavelengths as the Fraunhofer absorption lines seen in the spectrum of the photosphere. After all, they are caused by the same atoms, which both take up and give off energy at the well-defined wavelengths.

You can see this effect clearly by taking spectra as an eclipse begins. This technique was often used in the first half of the twentieth century but is rarely used any more. Donald Menzel became famous (among astronomers) in 1931 for his analysis of old eclipse observations of the spectrum of the chromosphere (Figure 3.11), and Menzel went on to make several sets of spectra of the chromosphere himself. One of the techniques was to move the film from top to bottom, given that the spectrum was spread out from side to side. In that way, you got a continuous photograph that showed how the spectrum changed over time. In particular, you can see that the brightest chromospheric lines turn into emission before the fainter lines do so. One of Menzel's most famous methods was the "jumping film" image, in which the film was moved very quickly between one exposure and the next, without a shutter.

As the photosphere wanes, its spectrum remains that of absorption lines, with only a couple of exceptions, such as for the strong H and K lines and for H-alpha. Then, for a few seconds, the absorption lines all across the spectrum turn into emission lines, as the H and K lines and H-alpha had a few seconds earlier. This emission spectrum seems to flash into view and then disappear again, and has the name flash spectrum. The flash spectrum appears because the bright photospheric Fraunhofer spectrum that overwhelmed it is covered by the Moon, and it disappears because the chromosphere itself is then covered by the Moon. A similar sequence appears in reverse at the end of totality.

3900

FIGURE 3.11   Moving film images made by W. W. Campbell during the 1905
eclipse in Spain: the film was moved vertically during the observation, the spectrum from a narrow strip of sun being exposed at any time, so that the vertical axis
shows how the spectrum changed over time at the onset of totality. Note K
(3,934), H (3,968), and H-epsilon (3,970).

The Discovery of Helium

People began taking spectra at eclipses in the 1860s. At the eclipse of
1868, P. J. C. Janssen of France was watching the flash spectrum with a
spectroscope—where "scope" instead of "graph" simply means that he
was looking through it with his eye. During the flash spectrum, he noticed several bright emission lines. One of them was yellow, so at first
he thought that it was the D line, one of Fraunhofer's original lines
then already known to actually be a pair of lines close together. This
pair of lines was known as D_1 and D_2. It resulted from sodium, and you
can reproduce these lines easily by throwing a pinch of salt into a flame
on your stove.

4000   4100

Janssen realized that this and some other lines of the flash spectrum were so bright that he might even be able to see them without an eclipse. He thus tried to do so the next day and succeeded in seeing them shine outside the solar limb. Almost simultaneously, in England, Norman Lockyer also tried and succeeded in seeing chromospheric spectral lines outside the solar limb with a spectroscope even without an eclipse.

With the ability to study the flash spectrum without an eclipse, Janssen and Lockyer realized that the yellow line they saw in the solar spectrum was slightly displaced from the D_1 and D_2 line. They thus called it D_3. Although the D_1 and D_2 lines were known to come from sodium, the D_3 line didn't seem to come from any known element. Thus Lockyer gave it the name "helium," from the Greek word "helios," the Sun. It wasn't until 1895 that helium was extracted chemically from other gases. We now know, of course, that helium is one of the fundamental elements, the second least massive nucleus after hydrogen. This discovery was perhaps the most significant ever to come from a solar eclipse.

Technically, the D_3 line is at 587.6 nm, while the D_1 and D_2 lines are at 589.6 and 589.0 nm. They are easy to tell apart when they both appear on a spectrum of sufficient quality, although the D_3 line is still hard to study. Helium needs the gas to be at a high temperature for it to produce this line, and the photosphere just isn't hot enough, so the helium D_3 line shows only weakly in the photospheric spectrum. It is therefore nearly impossible to make an image of the solar disk in this helium D_3 line. But another helium line exists at 1,083 nm, in the infrared and thus beyond the spectral region in which film is sensitive. Now helium at 1,083 nm is routinely mapped with telescopes on Earth using electronic detectors sensitive to the infrared, and you can call up such an image on the World Wide Web on most days (at http://umbra. nascom.nasa.gov/images/latest.html), if bad weather does not prevent the observing (Plate IV).

Imaging the Chromosphere

We mentioned earlier that we see into the Sun until the opacity is too great for us to see any farther. If we look in white light—all the Sun's light taken together—or in any continuous set of wavelengths, the limit of our vision is some height in the solar photosphere. Everything between it and us is pretty transparent.

But if we look at the wavelengths near the center of the strongest spectral lines typical of the Sun—mainly the calcium H and K lines and H-alpha—the Sun looks very different. The opacity at these wavelengths is very high because the light is strongly absorbed. Thus we don't see in as far. In fact, our line-of-sight stops somewhere in the chromosphere, and we can obtain an image of the chromosphere by putting a narrow passband filter in the optical path of the telescope.

It is not as easy to make such images of the chromosphere as it is to make images of the photosphere. We need to limit our vision to a very narrow band of wavelengths at and around the central wavelength of the strong emission lines. Normally, we want only 0.1 nm or so around the desired wavelength.

Traditionally, for the last hundred years, the device used to make such observations was the spectroheliograph—which literally translated means a spectrum-Sun-writer. In a spectroheliograph, the image of the Sun falls on a slit that extends across the Sun from top to bottom. Light coming through this slit is spread out in wavelength, usually so that color goes from side to side and position along the slit goes from top to bottom. In order to produce an image of the Sun in the desired narrow wavelength range, the trick is to have a second slit that passes only the solar light of whatever wavelength you choose. By making that second slit sufficiently narrow, you can allow as narrow a set of solar wavelengths to pass as you choose—as long as you lengthen your exposure time enough so that your film or other detector records enough light. This produces a spectrally pure image of the Sun, but it is only a long, thin piece, basically as wide as the slit and as high.

The further key trick is then to move the pair of spectroheliograph slits from side to side. The slits both move together, so that as the first slit scans the whole Sun from side to side, the second slit continues to pass the desired wavelength for each part of the Sun in turn. As a result of this scanning, an image of the Sun is mapped out in whatever wavelength was chosen. Often that wavelength is the H-alpha line or the calcium H or K line (Figure 3.12).

These spectroheliograms show the solar chromosphere. The stron gcr the absorption line, the higher the opacity and the higher up you see in the chromosphere. The H and K lines are the strongest, and thus show us the highest level of the chromosphere. H-alpha is not quite as strong, and so shows us a high but not quite as high level of the chromosphere. Thus by looking at various spectral lines, we can make a three-dimensional map of the chromosphere.

To see the chromosphere, you must isolate the emission line. Thus you can accept only a very narrow band of wavelengths. The problem with the method described above is that it is very slow, requiring up to an hour in some cases to produce an image of the entire Sun. In the 1930s in France, the astronomer Bernard Lyot worked out a way of making a filter that passed only such a narrow band of wavelengths. These filters used pure crystals of calcite, which have a property called

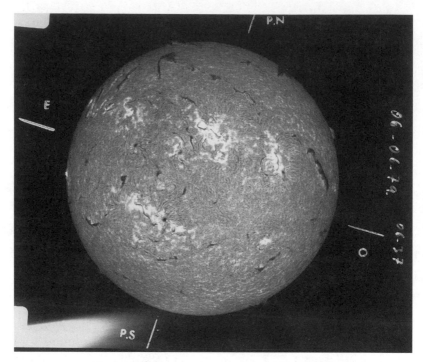

FIGURE 3.12   H-alpha (left) and calcium K line (right) images of the Sun,
6 June 1979.

birefringence, in which light that enters breaks into parts that are differently polarized, that are displaced from each other, and that travel
through the crystal at different speeds. Combining several crystals of
calcite of different lengths, Lyot succeeded in making the filters needed
to observe narrow wavelength bands. These filters are known as bire-
fringent filters or, often, Lyot filters. They have excellent optical quality and have what is known as square passbands: if you graph their
transparency against wavelength, the graph shows a constant level of
transparency in the central wavelengths with a constant level of
opaqueness outside it. Optically pure calcite is rare and expensive, and
is harder and harder to find, so that Lyot-type filters may cost $50,000
today. But they can give excellent solar images in H-alpha and in other
wavelengths, without the need for rastering or scanning across the disk.
Many Lyot filters were limited to a single wavelength each, though
some widely adjustable ones have also been built.

A much less expensive way of making a filter with a narrow passband is to deposit thin layers of reflecting material on a clear glass or quartz piece. The light bouncing back and forth between the deposited layers sets up a pattern of optical interference, and only a narrow passband emerges. These filters are much less expensive than Lyot-type filters, but their passbands are not as square. There is often more leakage of unwanted wavelengths, so the resultant image is not as purely from the desired spectral line. Still, these interference filters are so much less expensive and so readily available that they are widely used even by amateur astronomers. Both these filters and Lyot filters are usually temperature controlled, since the wavelength changes as the temperature of the filter changes.

Imaging the Sun in H-alpha or in helium in the visible spectrum shows the chromosphere, but there are more fundamental lines of hydrogen and helium that are in the ultraviolet, at wavelengths far too short to come through the Earth's atmosphere. Fortunately, special spacecraft outfitted to observe the Sun now observe these more fundamental lines on a daily basis.

The helium spectral line in the ultraviolet at 30.4 nm is one of the four lines observed with the Extreme-ultraviolet Imaging Telescope (EIT) on the Solar and Heliospheric Observatory (SOHO) (Plate V). The Lyman-alpha line of hydrogen, a spectral line similar to the red H-alpha line but from a more fundamental energy level of hydrogen and thus much stronger, is observed regularly by the Transition Region and Coronal Explorer (TRACE) spacecraft (Plate VI).

Appearance of the Chromosphere

The chromosphere as seen in H-alpha spectroheliograms or spectrograms shows a lot of structure that isn't seen in the photosphere. Sunspots themselves don't show well in H-alpha, but the regions around them are relatively bright. These regions are known as plage, from the French word for beach.

Within these plage regions in H-alpha we see narrow, dark filaments. These filaments turn out to trace out the boundaries between areas on the solar surface that are dominated by opposite magnetic polarities. When you see such a dark filament, it means that there is typically a large patch of outwardly-directed magnetic field on one side of the filament and inwardly-directed magnetic field on the other side. This implies that the magnetic field arches over the filament, and there is indeed good evidence that this is the case.

When the filaments rotate so that they are on the edge of the Sun, they stick up off the edge. We then see them emitting H-alpha, and call them prominences (Figure 3.13). Prominences can be very stable, and can last for months, though some do erupt in periods of hours. Although they stick up into the hot corona, they are made of cooler gas at chromospheric temperatures of 10,000 kelvins or so. Because they are both cool and stable, they are thus very different from solar flares, which are explosive phenomena at temperatures of millions of kelvins.

The calcium H line and K line images look very different from the hydrogen H-alpha images. Let us talk mainly about the K line images; the H line images look very similar, though they aren't always as pure when observed in filtergrams because the H-epsilon line of hydrogen

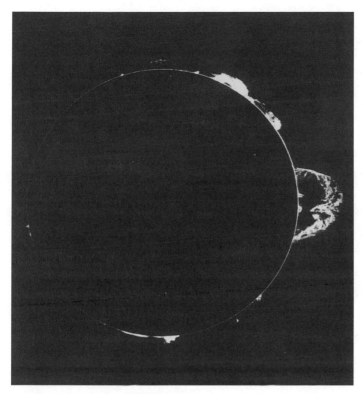

FIGURE 3.13   H-alpha image of a large prominence, 13 March 1970. The solar disc has been hidden artificially in the telescope.

has a wavelength very close to that of the H line of calcium, and so gets partially included in the images.

The K line images show polygonal regions that are the supergranules. Aligning them with magnetic field images taken of the photosphere shows that the supergranule boundaries, 30,000 kilometers or so across, match regions of moderately higher magnetic field. These magnetic fields at the supergranule boundaries are noticeably stronger than the average field in the middle of the supergranules, but still substantially lower in strength than the magnetic fields in sunspots. The boundaries are known as the chromospheric network.

If you look carefully at the limb of the Sun in H-alpha or some of the other strong spectral lines, you see that it is made up of little spikes. They have at various times been called "burning prairies" or "leaves of grass," and observers such as Father Angelo Secchi in the nineteenth

FIGURE 3.14   Drawing of spicules made in 1869 by the Italian Jesuit Father
Angelo Secchi.

century invested much effort in drawing them (Figure 3.14). Since the
mid-twentieth century, they have been called "spicules."

These spicules have been studied exhaustively, and can be seen to
rise and fall with periods of about 15 minutes. But the individual spicules are only about 1 or 2 arc seconds in diameter, so they are barely
visible and blur with neighboring spicules, especially because when we
view them on the limb each of our lines of sight no doubt goes through
several of them. There are hundreds of thousands of spicules on the
Sun at each moment, and there is still no general agreement about
what causes them.

The spicules consist of columns of chromospheric-temperature material moving upward, and all of them together carry enough matter to
replace the corona, the next higher level, in only a few minutes. Since
the corona is not doubled in density that often, we deduce that the
spicules cannot simply carry material only upward. Most of them must
fall back down, even if we can't see it in every case. Indeed, just watching spicules seem to fall doesn't prove that they are really falling, since
they could merely be becoming invisible at their tops as their chromospheric-temperature gas is heated up, and therefore would stop emitting in H-alpha. This example illustrates a general problem peculiar to
narrow-band spectral observations: something can seem to be moving
in your image either by changing temperature along its length, or by
actually moving. Only spectra, on which Doppler shifts can be measured, prove the case for the rising and falling of spicules by showing
the velocity of the moving material.

Some objects on the limb, seen especially in helium observations

FIGURE 3.15   TRACE coronal image taken on 10 July 1999, showing spicules
generally as dark absorbing features against a bright x-ray background at the limb.

from space, are larger than normal spicules and are known as macro-
spicules. They have been found to coincide with very small flares seen
in x-ray images and this might indicate that the spicules come from yet
smaller flares.

The TRACE spacecraft makes steady, high-resolution images of the
solar limb, in which many spicules are visible because the cool, dense
spicules absorb the extreme ultraviolet (EUV) light that TRACE is
imaging. Spicules are therefore detected in the EUV as dark features
(Figure 3.15) sticking up into the bright corona.

Spicules are important since the chromosphere seems to be made
entirely out of them. Although the chromosphere may appear like a
layer of gas, and has often been described as a layer and treated as
such in theoretical analyses, it is really a set of vertical spicules
each about 1,000 kilometers across and 10,000 kilometers high. If you
blur out the spicules and try to calculate temperatures, densities, and
other quantities for the chromosphere as if it were a smooth layer, you
get an answer, but your answer is not correct. Only by analyzing the

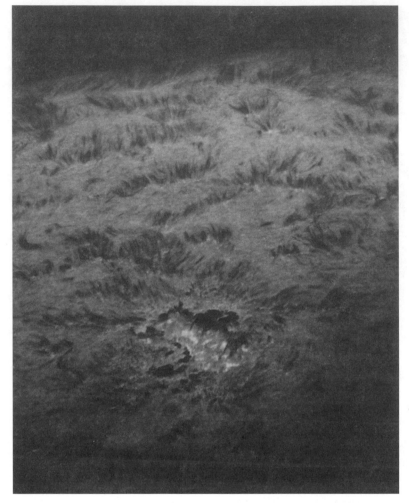

FIGURE 3.16   Spicule bushes seen in the wing of the H-alpha line near the limb of the Sun.

spicules individually do you actually find out what the chromospheric gas is like.

For a long time, astronomers tried to trace spicules seen in H-alpha down inside the solar limb to see what features they connect to on the disk of the Sun. The details of the connection are still unclear, but on the disk, not quite at the limb, one can see structures that look like bushes of spicules (Figure 3.16). These structures, no doubt, appear as spicules when they are seen at the limb. But these structures contain

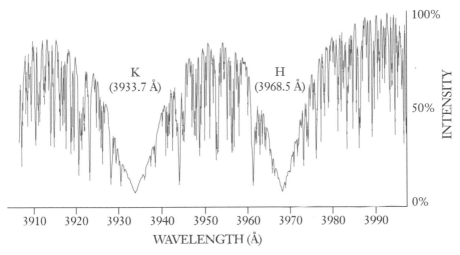

FIGURE 3.17 The solar spectrum in the vicinity of the calcium H and K lines. Narrow spectral lines are superimposed on the broad H and K lines.

both dark and light mottles, and there is not agreement as to which one corresponds to the spicules. The mottles seen on the disk often appear clumped in radial arrangements known as rosettes, perhaps indicating that they are associated with bundles of magnetic field lines sticking upward and spreading out from a small source.

Mapping the Three-dimensional Structure of the Chromosphere

We described above how the strong opacity in the center of the K line of ionized calcium makes our line-of-sight end high in the chromosphere. The K line has a complicated structure in wavelength (Figure 3.17) which can be useful in making observations at different heights in the chromosphere. Notice that from the continuous radiation (known as the continuum) that exists at both shorter and longer wavelengths on either side of the K line spectral structure, there is first a drop down in intensity toward the center of the line. On a finer scale than shows here, very close to the center, the curve reverses and rises back up. At the very center, it reverses again, resuming its downward trend. This structure is called a double reversal. At the very center of the K line, the intensity is only about 4 per cent of the intensity of the nearby continuum.

All of these reversals mean that the opacity varies for different wavelengths within the K line. Thus by looking at different wavelengths in the K line, you see down to different levels in the solar atmosphere. In the center of the K line, known as K_3, the opacity is high, and your view terminates at the highest level of the chromosphere. As you go outward to the emission peaks near K_3, which are known as K_2, the opacity is somewhat lower and you see farther into the solar atmosphere, into a level lower in the chromosphere. As you go through the dips, known as K_1, and farther outward until you reach the continuum, you see still lower and lower levels. By the time you are at the continuum, you are seeing the photosphere. These different levels show in spectroheliograms taken in different parts of the K line (Figure 3.18).

Some astronomers spend their time calculating the details of the Sun based on studies of spectral lines, a field of study known as "stellar atmospheres." It is easy to see how as you go inward toward the line center from the continuum, you are looking through gas at higher opacity, and so you are seeing to levels higher than the photosphere from which you start. What is less obvious is that when you get to high levels of the chromosphere, collisions in the gas at some level in the chromosphere cause the K_2 peaks to form in the spectrum. At still higher levels, some of the photons of light that are given off can escape readily to the outside. This escape gives you the double reversal in the center of the K line. The calculations involved in understanding how light moving through the solar atmosphere interacts with the gas atoms in that atmosphere—a field of study known as radiative transfer—are often made with some of the most powerful supercomputers available. They often assume that the Sun is made of layers; in the calculations, the temperature and density at the top of one layer are forced to match the temperature and density of the bottom of the next higher layer, as must be the case physically, and the layers are progressively stacked on top of each other to build up a simulated atmosphere. Although we know that this type of model is wrong for the atmosphere as a whole, it might be a good approach to modeling the properties within each type of vertical structure.

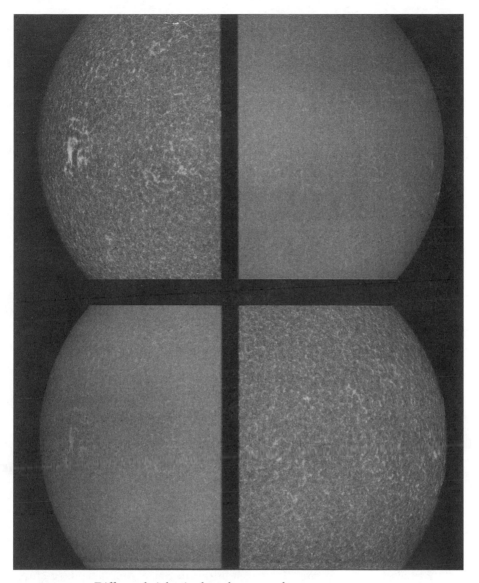

FIGURE 3.18   Different heights in the solar atmosphere are seen as we tune
through the K line. These four spectroheliograms were taken in K_3 (upper left
and lower right) and K_1 (lower left and upper right).

But the radiative-transfer explanation is not the only one. If you take spectra at extremely high spectral and spatial resolution, you can see that the peaks seem to be concentrated at wavelengths slightly to the left or to the right of the K_3 central wavelength. Phenomenologically, one can simply say that these features, which presumably correspond to spicules, are Doppler shifted to the left or to the right because of their motion toward or away from us. The K_3 spectral feature can be caused by a cooler gas with low velocity that surrounds the spicules. Such a detailed look at the fine structure gives an insight that is not available from lower-resolution observations. Again, the structure of the chromosphere that you derive from the average blurred image is different from the actual temperature, density, and velocity structure of an individual spicule or even of an average of the values for many spicules. For example, the height of the chromosphere is given, on the basis of average values, as 2,000 kilometers above its base, while we clearly see spicules extending to over 10,000 kilometers high.

In the visible spectrum of the Sun, only the H and K lines of ionized calcium are strong enough to show the double reversal. The hydrogen lines, even H-alpha, are not that strong. But as you go to the ultraviolet, there is a pair of ionized magnesium lines at wavelengths just beyond the wavelengths that come through the Earth's atmosphere. When they were first studied well from Orbiting Solar Observatory 8 in the late 1970s, these ionized magnesium lines were given the names h and k, by analogy with H and K of ionized calcium. Historically, of course, it is a very bad notation, since it would have made more sense to extend Fraunhofer's notation to, say, P and Q rather than repeating the alphabet in lower case. But the h and k notation seems to have stuck (Figure 3.19).

It is particularly important to understand the double reversal in the structure of the calcium K line because such double reversals are observed in the K lines of many distant stars. Indeed, the presence of such K lines is thought to indicate that these distant stars have chromospheres, like the Sun in type but much more powerful. The Sun's chromosphere is so puny that it would not be detected at all if the Sun were

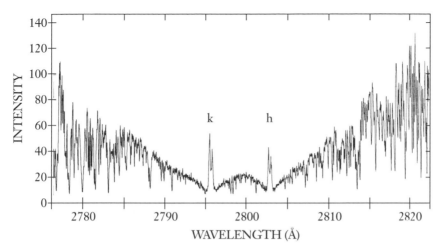

FIGURE 3.19   The h and k lines of magnesium farther into the ultraviolet than the human eye can see. The letters were assigned by analogy to the H and K lines of calcium, so the h line is at a longer wavelength than the k line.

as far away as these stars are. The strength of the K line K_2 reversals varies on the Sun with the solar activity cycle. Long-term studies of many stars, especially those made at the Mt. Wilson Observatory, have found variations of the K_2 reversals with periods of several years. These stars presumably have sunspot cycles that are causing these reversals.

In the 1950s, Olin Wilson of the Mt. Wilson Observatory and M. K. Vainu Bappu, an Indian scientist who often worked in the United States, discovered that the width—not the strength—of the central reversal correlated very well with the brightness of the star. Astronomers are always eager to have ways to find out how far away stars are, and this discovery provides a tool for determining distance: by comparing the absolute brightness of the star with its brightness as observed here at Earth, we can calculate how far away it must be. The link between the width of the K_2 line and brightness is known as the Wilson-Bappu effect (Figure 3.20).

Prominences

Prominences are phenomena that vary with the sunspot cycle, and so are considered to be part of the active Sun. The chromosphere as a

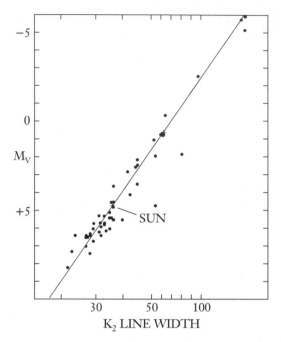

FIGURE 3.20   The strong correlation between the absolute brightness of a star
and the width of the K₂ line in the spectrum known as the Wilson-Bappu effect.

whole, in contrast, is always there and so is part of the quiet Sun. But
prominences are similar in temperature and density to the chromosphere, so they can be thought of as extensions of chromospheric gas
higher into space.

Prominences come in many shapes and types, and detailed classifica-
tion schemes have been worked out for them. A large, long-lasting
type is a hedgerow prominence, and such quiescent prominences extend for perhaps a hundred thousand kilometers along the edge of the
Sun (Figure 3.21). Though we may lose track of it when the rotation of
the Sun carries it over the limb and out of sight, it may reappear 2
weeks later on the Sun's opposite limb and then disappear as it moves
onto the face of the disk, and then reappear again on the original limb
another 2 weeks later.

Other prominences are of the eruptive type. If you watch the Sun
through an H-alpha filter for a few hours, you will see these prominences change, rising and falling and sometimes blowing off the Sun into

FIGURE 3.21   A hedgerow prominence.

space. Such events may take many minutes or even hours to unfold, and they are not immediately apparent to the eye. To study these events, prominences are sometimes viewed in time-lapse movies, in which 1 frame is taken every minute and then the movie is played back at 30 frames per second. Thus the motion is speeded up by a factor of 1800, and 30 minutes of time passes by in 1 second on the screen. Ten seconds, then, shows 300 minutes or 5 hours of prominence motion, and you see prominences rise, fall, seethe in place, or erupt.

Prominence eruptions are often seen to occur simultaneously with solar flares, but whether or not prominences erupt because of flares is still under debate. Also under debate is the link between eruptive prominences and coronal mass ejections (CMEs), large-scale ejections of hot coronal gas. Such eruptions are now studied much better than previously because of the wide field of view of the Large Angle and Spectrographic Coronagraph (LASCO) on the SOHO satellite.

Spectra taken of prominences show that they have very complicated

FIGURE 3.22   A spectrum obtained in a prominence shows strong evidence of
motions: the image would be a straight vertical bar if there were no Doppler shifts
due to motions in the prominence toward and away from the viewer.

motions (Figure 3.22). In Figure 3.22 the vertical direction corresponds to position along the prominence, while wavelength goes from
left to right. The white regions show the prominence gas, and whether
they are at the left side or right side of the center shows whether their
wavelengths are blueshifted (to the left) or redshifted (to the right).
The figure shows that the prominence contains gas that is moving rapidly, with both large blueshifts and large redshifts, and with a large
range of motions at a given vertical position.

The prominences show particularly well in the helium lines mapped
by the Skylab Apollo Telescope Mount in the 1970s and, more recently,
from SOHO. In the Skylab images, the ultraviolet wavelength of helium
(30.4 nm) overlaps slightly with some coronal images but is basically an
isolated emission line. So such images can be taken with spectrographs
which form images of the Sun that are positioned alongside each other

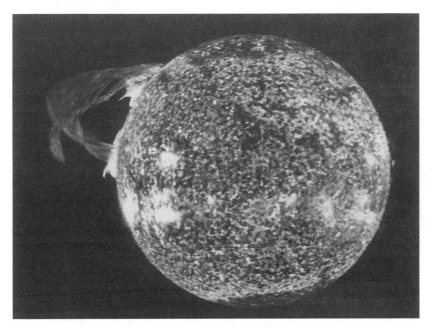

FIGURE 3.23   Skylab image of the Sun in the singly-ionized helium spectral line at 30.4 nanometers showing a prominence eruption.

according to wavelength, sometimes overlapping each other when the images are from emission lines having nearly the same wavelength. For the case of the strong helium line you can for the most part, just cut out the helium part of the image and ignore the rest (Figure 3.23).

THE TRANSITION REGION

Between the relatively cool photosphere and the hot corona one finds not only the chromosphere, but also a part of the atmosphere that covers the range from the 10,000 K (roughly 20,000°F) temperature of the chromosphere, to the several million degrees of the corona. This part of the solar atmosphere is usually called the transition region. It does not receive as much attention as the other parts of the atmosphere, and is often thought of merely as an interface between the chromosphere and the corona. However, it is a highly dynamic place, which responds readily to local changes in the heating rate. This makes it ideally suited

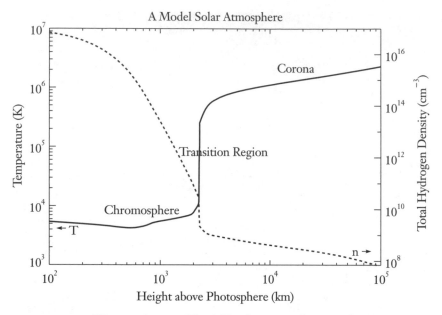

FIGURE 3.24   The temperature (T) of the solar atmosphere as a function of height above the photosphere, for a plane-parallel atmosphere, with no structure in the horizontal direction. The extremely thin region containing the rise from 10,000 to 1,000,000 degrees is called the transition region. The number density (n) of hydrogen is also shown.

for studies of the coronal heating mechanisms, and for searches of the source of the outflow of material known as the solar wind. According to atmospheric models, such as Figure 3.24, the transition to coronal temperatures happens extremely quickly, over a very short distance.

Figure 3.24 is misleading in at least one respect. This type of display treats the solar atmosphere as if it were a uniform layer, one that changes only as a function of height, but is uniform in the horizontal direction. This is decidedly not the case, as is clear from the photographs we have included in this chapter. The atmosphere is actually highly nonhomogeneous, and the temperature is very poorly correlated with height. In particular, we are finding that material at chromospheric and transition region temperatures can be found at great heights above the surface. This material has a complicated fine structure and is often in a highly dynamic state of motion.

A recent article on the transition region listed the following basic questions that remain to be answered:

1. What is the structure of the atmosphere in the transition region, and what is its relationship to both hotter and cooler structures?

2. Why does the transition region seem to be mostly empty space—why are the structures so small and why do they fill only a small fraction of the volume?

3. How do we explain the distribution of hot and cool material in the transition region, especially the increasing amounts of cool material below 100,000 K, after a decrease is seen between 1,000,000 K and 100,000 K?

4. Is the average model, such as Figure 3.24, an accurate picture of the actual structures in the transition region?

5. What is the heating mechanism that keeps the transition region in its observed state?

These questions have not yet been answered, largely because the transition region is very hard to observe. New and better observations, mainly from space, will be needed for progress to be made. We discuss some of the plans for new observations in Chapter 6.

What We Don't See

From the evidence brought to light by research in archaeo-astronomy, it seems that humans have been constructing instruments to supplement their sensory equipment for many thousands of years. Stone markers, crude sighting devices, and methods for keeping track of monthly and seasonal events were in common use worldwide. The culmination of these naked-eye observations was Tycho Brahe's Urani-borg Observatory—located on an island near Elsinore castle in Den-mark—from which Tycho and his assistants obtained planetary orbit determinations near the end of the sixteenth century so accurate that Johannes Kepler was finally able to figure out the true shape (elliptical) of the planetary orbits.

But it was Galileo's use of the telescope a few years later that brought home in dramatic fashion the realization that there are strange and wonderful phenomena in the heavens that we cannot see with the na-ked eye. The phases of Venus, craters on the Moon, moons circling the planet Jupiter, and sunspots on the face of the Sun were among the new discoveries revealed by this instrumental extension of the human appa-ratus.

We have been constructing bigger and better telescopes ever since, continuing right up to the present day. But we have also discovered that restricting our attention to the wavelengths that our eyes can detect is a major limitation. Astronomical objects in general, and the Sun in particular, look markedly different at radio, infrared, visible, ultraviolet, and x-ray wavelengths. By exploring those differences we can begin to understand what it means for the Sun or a star to look so different in x-rays or radio than in visible light.

THE NONVISIBLE

The light that we do see is only a small fraction of the total emitted by astronomical objects. Most wavelengths of light are not able to penetrate our atmosphere to reach us down at ground level, as shown in Figure 4.1. Short-wavelength ultraviolet (UV) and x-ray light are absorbed before reaching ground level, as is infrared (IR) light. It is surely no accident that the light that reaches us through the atmosphere happens to coincide with the light we use for seeing; our eyes adapted to work with the light available (although it is not at all obvious that our atmosphere had to be transparent at the dominant wavelengths emitted by the Sun). The objects that we want to study— the Sun, planets, stars, galaxies, black holes—do not know anything about our atmosphere and much of their energy is emitted at wavelengths that do not survive to ground level. The only way we can detect these wavelengths is to put our equipment above the Earth's atmosphere.

THE SOLAR CORONA

Above the visible surface of the Sun a strange thing happens, something that puzzled astronomers for more than 50 years. Imagine the following: You turn on the burner of your stove, and then explore the air just above it. As you might expect, just above the surface of the burner the air is hot, but not quite as hot as the burner itself. You probe a fraction of an inch higher and the air has cooled a bit more, although it is still hot. So far, all is well, in that the burner is heating the space around it, but not to a temperature higher than its own. But then you

FIGURE 4.1 Transmission of the Earth's atmosphere for wavelengths of light from x-ray to submillimeter waves; radio waves, of millimeter wavelengths and longer, extend to the right of the graph. The curve shows the altitude at which half of the incoming light is absorbed by the atmosphere. Note that only a few small "windows," in the visible and near-infrared, exist in which the incoming light reaches ground level. Almost the whole radio spectrum also reaches the ground through a window of transparency.

move another fraction of an inch higher and suddenly the temperature shoots up, becoming 10 times as hot as the burner.

This unlikely situation is what we find in the atmosphere above the visible surface of the Sun. The corona is very tenuous and would qualify as a very good vacuum in any Earth-based laboratory, but it is very hot, reaching temperatures higher than 5,000,000°F only a short distance above the 10,000°F surface. The bad news, from the viewpoint of an observer, is that the extremely high temperature of the corona means that the radiation it emits is not ordinary visible light, but x-rays. As an object heats up, the light it emits moves from the infrared, to the red, then into the violet. In the corona, the temperature is so high that the average energy of a photon of light emitted by the gas is increased so much that the "color" has been shifted far into the blue, into the ultraviolet and extreme ultraviolet. A small amount of visible photospheric light, scattered from the coronal gas, makes the corona visible under some circumstances, but the actual emission from the corona is mainly at short wavelengths.

In Figure 1.11 we showed a ground-based, visible-light image of a total eclipse. That same eclipse, with the hot corona inserted into the picture, is shown in Plate VII. The inner portion shows emission from the million-degree corona, obtained from a telescope that images the extreme ultraviolet emission of this hot gas directly.

Even if these short wavelengths did reach the ground, we would not see them because our eyes are blind to those colors. So the difficulty is dual, because we need to build instruments that can see colors that humans normally cannot see, and then we have to put those instruments up above the atmosphere and somehow record the images and get them back down to the ground. Some of the special difficulties encountered in putting instruments in space are discussed in Chapter 6.

The advantages of looking at the Sun in these invisible wavelengths is clear, even from a single "snapshot" view. Plate VIII shows a solar active region viewed in the emission of 9 times ionized iron, resulting from gas so hot that electrons crash into iron atoms forcefully enough to knock the first 9 of their 26 electrons completely out of the atoms. Such highly ionized atoms are present only when the coronal gas is at a temperature of about 1,000,000 K. This type of image cannot be obtained from the ground, and reveals a Sun that is markedly different from the bland white-light photosphere.

The explanation for this complex, hot, and dynamic corona still eludes us after nearly 60 years of effort. It is almost certainly the case that a central feature of the answer lies in the magnetic fields that are generated inside the Sun and then bubble up to the surface. This explanation is unfortunate for scientists trying to explain the corona in simple terms, because most people do not have any intuitive feeling for the properties of magnetic fields. In addition, the high temperature of the corona means that the coronal gas is ionized—the electrons of its atoms are mostly separated from the nuclei—and most humans have no significant experience of this fourth state of matter.

Plasmas, hot gases in which a substantial number of the atoms are ionized, are inherently complex, and magnetized plasmas are enormously more complex. Moreover, the magnetic field is anchored in the relatively dense solar surface, where the motions of the solar matter are strong enough to affect the field, and it reaches out into the corona, where the coronal matter density is so low that the magnetic field dominates and constrains the gas. The physics of these two regions are quite different, but they are connected to each other by the fact that the field threads from one to the other, and the gas can flow along the field from one region to the other. It is therefore not too surprising that we have been making only slow progress in understanding the solar corona.

SOLAR FLARES

The corona has long been thought of as a hot but placid plasma. Only in recent years, especially with observations of daily coronal mass ejections with the Solar and Heliospheric Observatory in space, has the corona's dynamic nature become apparent. But even these events do not compare with the violent nature of a solar flare.

On the cosmic scale, solar flares are not very big events. A large flare might increase the total energy output of the Sun 0.1 per cent for a few minutes, while other solar-like stars in our galaxy produce flares a thousand times larger. At the extreme, there are believed to be explosions— perhaps due to the interaction of two black holes—that liberate in one second more energy than the Sun will emit in 10 billion years. But on our earthly scale, the solar events are enormous, each with enough energy to power all human activities for centuries.

Even though we are more than 90 million miles away, we feel the effects of solar flares. Radio communications are disrupted—one of the first clear indications came during World War II, when a British researcher worried that the Germans had found a way to jam the newly invented radar, though the Sun turned out to be the culprit. Power distribution networks have been damaged, satellites have been put out of commission, and the radiation from a flare can be fatal to unprotected astronauts. These, and many other flare effects, have been studied extensively, and many of them are described in more detail in Chapter 8 of this book.

But what do we know about the causes of these eruptions on the Sun? Why does the Sun produce these sporadic outbursts? Equally important, why doesn't it produce more and bigger eruptions?

Unfortunately, there are nearly as many flare theories as there are flare theorists, since many different models are compatible with the laws of physics, and the observations are not yet able to rule out all of the possibilities. But progress is being made, thanks to both improved observations and improved theory, and at least a broad consensus is emerging. Most flare researchers agree that the magnetic fields that erupt through the solar surface into the corona are somehow involved. The reason for this agreement, aside from some observational clues, is that an explanation of flares requires some type of energy storage and a mechanism for rapid release of this stored energy. Magnetic fields can provide both, and we also observe drastic changes in the magnetic field during a flare. Figure 4.2 gives one of the most frequently invoked magnetic field configurations, which we will call the Standard Model.

What this figure shows is an arch-shaped array of magnetic field lines that are supporting a prominence, a blob of cool, dense material. (For historical reasons, when a prominence is viewed on the disk of the Sun it is called a filament. When it projects up above the limb and is seen against the dark background of space, it appears bright, and is called a prominence; when it is viewed face-on against the bright disk of the Sun, it appears relatively dark and is called a filament.) At the top of this arcade the magnetic field is weak and the field lines open up to interplanetary space, forming a shape known as a helmet streamer (because of the similarity to the tops of Prussian military helmets). These structures are seen often in the corona, usually in or near active re-

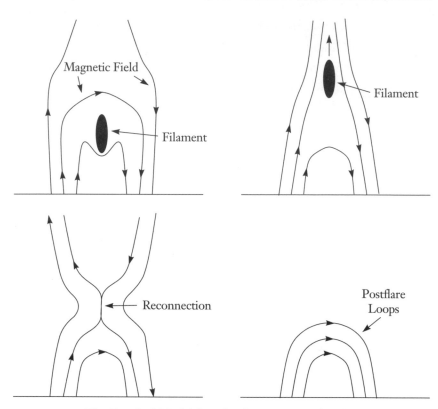

FIGURE 4.2   The Standard Model for solar flares. A mass of cool, dense material being supported by an arcade of magnetic field lines becomes unstable and erupts outward, breaking open the field. Reconnection of the field lines to their closed state releases energy to produce the observed brightenings.

gions, but there are also larger and smaller replicas of this configuration found throughout the corona.

The situation shown in the top left panel of the figure is dynamic, in that the magnetic fields slowly evolve, and the prominence material flows and moves in all directions, but on the whole it is stable for many hours or days at a time. Occasionally, however, for reasons that are not yet understood (and therefore cause passionate debate among solar researchers) the filament is flung upward and an enormous amount of energy is released, as shown in the next two panels. The magnetic field is broken open, and then connects back underneath the departed filament, and the reconnected "postflare" loops glow brightly at millions of degrees for many minutes or hours. Usually, the region forms itself back into the initial configuration, including the filament, very soon after the flare.

Recent observations from the TRACE satellite, part of NASA's Explorer program, show that the Standard Model works well as an explanation for large flares. Plate IX shows the development of a flare. In the first panel, the flare is just beginning: part of the filament can be seen erupting upward at the left of the frame, and bright ribbons running horizontally across the middle of the frame outline the feet of the arcade of magnetic field lines.

In the subsequent frames, the postflare loops start to form. This particular flare has a second filament eruption occurring in the background—it can be seen moving upward by comparing the first and third frames—and a second set of postflare loops is visible in the last frame. In this example, the filament and the arcade of magnetic field lines supporting it actually took on a circular, rather than a linear, shape. The postflare loops therefore have a circular configuration, like a bent tunnel. But the principle of magnetic field opening and reconnection is followed in this event.

LOOKING INSIDE THE SUN

Most of the phenomena we observe at the Sun's surface have their origin in the solar interior. Here we will discuss two of the major problems in modern solar interior research: the core of the Sun, where nuclear fusion takes place, and the outer convective envelope, where magnetic fields are generated.

The energy source of the Sun, nuclear fusion, is located deep in the center where temperature and pressure are high enough to ignite nuclear burning. We have very good models of the solar interior, and one result of these calculations is that the nuclear fusion that powers the Sun should also produce a certain number of subatomic particles called neutrinos. But experiments carried out here on Earth do not detect as many of these neutrinos as are predicted. So there is a problem, usually referred to as the neutrino problem.

The location of magnetic field generation is difficult to pin down:

it may be located anywhere from a shallow surface layer just below the photosphere down to an hypothesized boundary layer between the outer convective envelope of the Sun and a nonconvective (radiative) core. Both cases are possible, and finding the right answer involves a delicate interplay between observations and theory that can serve as a model for the way in which science operates.

An aside on how science operates: The key breakthrough that made the Scientific Revolution possible at the end of the sixteenth century was the realization that having a plausible theoretical explanation is not enough, and that a theory must be tested against the actual observed conditions. The famous moment when Kepler realized that the observed position of Mars differed by 8 minutes of arc from the position calculated with circular orbits is a prototypical example. These ideas about the structure of science mean that we need to come up with a theory that not only sounds right, but also actually works.

Explaining magnetic field generation inside the Sun requires a knowledge of the fluid flow patterns below the solar surface. Only a few years ago it was thought that such knowledge would remain forever out of observational reach, limited strictly to theoretical calculations. However, a key realization—that waves permeate the Sun and can be analyzed by careful study of oscillations at the solar surface—opened up an entire field of study, helioseismology, that is revealing to us the once-hidden state of the interior.

Neutrinos

One of the great joys of studying science is the occasional rare moment when you are struck by a bit of understanding that completely changes your view of the world. Studying the behavior of neutrinos offers such an opportunity, since they are quite outside of our ordinary experience.

Neutrinos were first postulated on theoretical grounds by Wolfgang Pauli, in order to explain certain nuclear decay processes—in which one type of nuclear particle changes into another type—that did not seem to conserve mass and energy or a property of nuclear particles called spin. For instance, a neutron just sitting by itself alone in empty space will spontaneously change into a proton plus an electron. The problem is that the total mass and energy of the observed decay products do not equal the initial amounts of these quantities, and the spin seems to change as well. More subtly, the way in which the directions and speeds of the decay products (the proton and the electron) are distributed in space seems to violate the mechanical law known as conservation of momentum, unless an additional undetected particle is being produced.

The experiments that Pauli was studying did not show any other particles coming out, but he was not ready to give up on the important laws conserving energy and momentum. So he proposed the "little neutral one," or "neutrino," a particle that is emitted in these decay events, but that, for some reason, we do not detect. It carries mass, energy, momentum, and spin. The neutrino remained only a theoretical construct for many years: Pauli proposed the existence of these particles in 1930, and it was the mid-1950s before Clyde Cowan and Fred Reines were able to detect their presence (and the 1990s before a Nobel Prize was awarded for their discovery). These neutrinos are now called electron neutrinos, since they are considered to be members of a family of particles that contain electrons along with the "up" and "down" quarks. A second family of particles was later discovered containing a muon (a type of heavy electron), the "strange" and "charm" quarks, and a corresponding neutrino, the muon neutrino. A third family contains a still heavier type of electron known as the tau particle, the "top" and "bottom" quarks, and the corresponding tau neutrino. This third type of neutrino wasn't actually detected until 2000, when signs of its existence were observed in the giant atom smasher at Fermilab, near Chicago.

So the neutrino does in fact exist, and the reason that it is so hard to see is that it does not interact with matter via the electromagnetic force. This plain statement is the one that is at first hard to comprehend: in order for matter to be detectable—in order for it to have visible effects—it must be a "carrier" of a force that we can detect. Put another way, one bit of matter cannot interact with another unless it has within itself the means for interacting. This platitudinous-sounding statement has real implications, because it means that if the particle does not carry the necessary type of force, then it does not interact via that force.

Astoundingly, it seems that there actually are types of matter that do not interact (or hardly interact, to be more precise) with our sensory or observing equipment. This means that they do not affect us or the world around us as they fly by, and this in turn means that we have no way of knowing that they are there. This is a serious problem for science, because it does no good to hypothesize unless there is some way to test the hypothesis; proposing the existence of the neutrino does no good unless there are ways to check whether neutrinos actually exist.

Neutrinos do interact slightly with matter, via a force called the weak nuclear force. Its name indicates in part that this interaction is astoundingly weak compared to the electromagnetic force that accounts for almost everything we see around us. For example, roughly 100 trillion neutrinos per second from the Sun pass through your body, but the probability of just one of these actually interacting each *year* with one of your atoms is less than 0.0001 per cent—which is why you don't notice them.

Given how hard it is to detect neutrinos—stopping a typical solar neutrino requires about 1 light year (5 trillion miles) of lead—it would seem to be a practical impossibility to build an experiment to detect them. But one should never discount the cleverness of experimenters, and ways have indeed been found to measure at least a few of the solar neutrinos that are flying past us from the Sun. The longest-running experiment is that of Ray Davis, placed deep in the Homestake mine in Lead, South Dakota, and usually referred to as the Homestake experiment. In this setup 100,000 gallons of carbon tetrachloride (cleaning fluid) are used, to take advantage of a reaction in which a neutrino gets absorbed in a chlorine nucleus, converting it to an atom of argon as an electron is emitted. The emission of the electron conserves electric charge (and a few other quantum numbers).

The most astounding part of this experiment is that Davis found a way, via chemical analysis, to detect that 1 atom of argon in this enormous vat of material. Every few months the fluid is sent through the detection equipment and the argon atoms are counted, so that all of the neutrino events during those few months are summed up as a single total number. This is not, therefore, a "real-time" detector, since it might be months after an event before one knows that it occurred. But it is nonetheless an impressive technical achievement.

The main problem with this experiment is that the neutrinos must have a high enough energy to produce the chlorine-to-argon reaction. It requires 0.8 million electron volts (MeV), whereas the neutrinos produced by the proton-proton chain—the dominant reaction in the Sun, building up helium by adding protons one by one to an initial proton—have a maximum energy of 0.4 MeV. The Homestake experiment is therefore sensitive only to the much rarer higher-energy neutrinos produced from some of the other less common fusion reactions in the Sun.

The same is true of another experiment, the Kamiokande, and its bigger brother, the Super-K, which rely on collisions between incoming neutrinos and electrons in a large batch of ultra-pure water. The interaction is detected by placing an enormous number of light-sensitive photomultiplier tubes around the interaction region (Figure 4.3). The advantage of this experiment is that it detects the events as they happen, and it can also provide information about the direction from which the detected neutrino arrived. A new experiment, BOREXINO, located in a chamber adjacent to the automobile tunnel passing through the Gran Sasso mountains in Italy, uses the scattering of incoming neutrinos from the electrons in a large vat of ultra-pure liquid scintillator surrounded by thousands of photomultiplier tubes. The experiment will be far more sensitive than previous ones and will also provide information on the neutrino energy as well as real-time detection of the neutrinos.

Experiments set up in Russia and Italy, known as SAGE and GALLEX, use a neutrino reaction in the element gallium, converting it to germanium. The process is similar to the chlorine reaction, but it

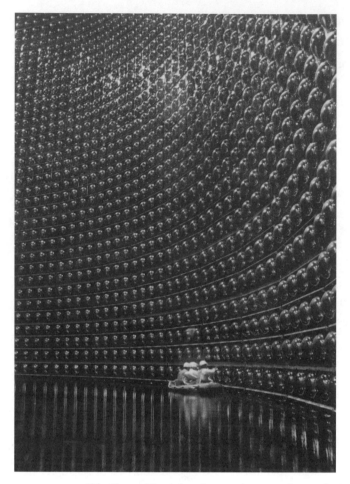

FIGURE 4.3   The Super-Kamiokande neutrino experiment during its first filling
with ultra-pure water. Most of these huge light-sensitive detectors exploded in
2002 when the tank was being refilled.

does not need a neutrino energy nearly as high. The reaction requires a
minimum neutrino energy of 0.2 MeV, so that it is sensitive to the
dominant proton-proton neutrinos, and the expected neutrino detection rate is some 20 times higher than that of the other experiments.
Despite the enormous difficulty and expense of amassing tons of gallium and analyzing it for the occasional germanium atom, these experiments have also run successfully.

The Sudbury Neutrino Observatory (SNO), deep in a mine in Sudbury, Ontario, contains a huge container of heavy water—water in which deuterium is substituted for ordinary hydrogen: D_2O. SNO is sensitive to all flavors of neutrinos, not only the type recorded by the Homestake and Kamiokande experiments, so it should be crucial in resolving whether neutrinos change from one flavor to another, important for such studies. SNO is now collecting data.

In all of these experiments the answer is the same: not enough neutrinos are detected to match the predictions. Typically, each experiment detects only about half as many as should be seen. Originally, when first the Homestake and then the Kamiokande results became known, it was thought that the theory could be adjusted, since those experiments are sensitive mainly to much rarer solar reactions (although some details of the comparison between Homestake and Kamiokande would still be very hard to explain). With the new experiments, the situation has only deteriorated, since the shortage of low-energy neutrinos—those produced by the main fusion reactions and therefore tied directly to the total energy output—is much harder to explain.

This persistent shortage of solar neutrinos, the neutrino problem, is a "problem" only in the sense that there is no consensus as to the cause. There are many possible explanations, ranging from the prosaic to the wildly exotic. For instance, models that try to explain the low neutrino rate via something about the Sun generally seek to cool the interior, so that the balance of nuclear fusion processes is such that the same total energy is produced but the number of neutrinos is lower. Such solutions invariably run into the problem that the models of the solar interior are so good that most of these proposed answers can be ruled out.

One of the more exotic proposals, which is widely believed to hold the answer, involves "neutrino oscillations," in which a quantum-mechanical effect (named MSW after the scientists who proposed it: Lincoln Wolfenstein, Nikolay Vladimirovich Mikheev, and Alexi Smirnov) changes the neutrinos from the type that would be detected into another type, which would not have been detected with the existing experiments. Only better data, from the new experiments about to be done, will help to resolve this question.

Helioseismology

Any modern geology textbook will have a description of the Earth's interior, even though direct exploration of the layers beneath the Earth's surface has reached a depth of only a fraction of 1 per cent of the Earth's diameter. How then is it possible for us to know the composition and structure of the interior?

One of the tools used in such studies is seismology. When a sudden large release of energy occurs in a localized spot near the surface—such as an earthquake or an atomic bomb blast—the energy release causes a deformation of the Earth's crust that results in waves propagating outward, much in the manner of a hammer striking a gong. Some of these waves travel along the surface, while some move down into the interior, gradually spreading and weakening as they move away from the source.

Sensitive motion-detection apparatus located thousands of miles away can detect these disturbances, and can be used to determine the location and strength of the event. More important for our purposes, detailed analysis of the incoming waves at the detection site can provide information about the composition of the Earth in between the source and the detector. Waves that started out moving downward into the Earth are often deflected by boundaries or changes in composition, and may be bent back up to the surface. In this way it is possible through seismology to study the inner portions of the Earth that are not otherwise accessible. For instance, a certain type of wave called a shear wave cannot propagate through a liquid. So if a comparison of the waves reaching the detector via the Earth's surface with waves arriving after having traveled below the surface shows that shear waves are missing from the latter, then we know that the waves traveled through molten or liquid material on their way from the source to the detector.

In order to have wave motion, two things are required: a disturbance and a restoring force. For instance, a block of gelatin sitting on your dessert plate has an equilibrium shape, more or less like a cube. If you poke it with your fork, you displace it away from its relaxed shape (this is the disturbance), while the attraction of the gelatin molecules to each other tends to restore the cube to its original shape. The result is that waves travel rapidly up and down the slab of gelatin, which appears to our slow eyes to be vibrating back and forth. A more traditional example is the plucked violin string. The string is stretched tightly between two attachment points, so that there is a tension force along the length of the string, which tends to keep it straight between the two ends. If the string is disturbed sideways (plucked), the tension will bring it back toward its preferred straight shape.

The alert reader might at this point ask why the gelatin or violin string goes through many oscillations, instead of merely returning directly to the undisturbed state. The answer is inertia. The gelatin or the string has mass, so that as it swings back toward the neutral condition, the speed of the moving matter pushes it right through the equilibrium position. The behavior is like a pendulum, which after having been displaced and released will swing through the vertical and up the other side, then back again. In principle, any of these oscillations would go on forever, unless the motions are eventually damped out. For instance, if the above experiments are carried out in a large vat of honey, it is clear that the gelatin or the violin string would quickly stop vibrating.

The surface of the Sun is seen to have oscillations, rather like the surface of a swimming pool. These oscillations turn out to be due to the combined effect of an enormous number of different sound waves running across and into the Sun, traveling in all directions and with a large range of sizes (or wavelengths); it is not just the surface, but the interior as well that is oscillating, as shown in Plate XA. Most important for our purposes is that some of these waves penetrate deeply into the Sun and then return to the surface, as shown in Plate XB. The waves reflect back from the surface because of the abrupt upper boundary at the solar surface, so they go down again, are bent up again, and so on. A wave may make many such bounces as it travels around the Sun.

What all of this means is that waves of various frequencies (and various wavelengths) are reflecting up and down between the surface and a certain depth as they travel around inside the Sun. By very carefully measuring the properties of oscillations at the surface, which we can

see, we therefore ought in principle to be able to determine some properties of the solar interior, which we cannot see.

The longest wavelengths—those having the lowest frequency of oscillation—penetrate most deeply inside the Sun. If we want to study the
deep interior, we therefore need to have uninterrupted observations for
a long period of time, in order to record enough full oscillations to
get an accurate measurement. Unfortunately, we have day/night cycles
at any given spot on the Earth, while measurements lasting several
months or longer are needed. Several options are available. We can set
up equipment at the South Pole during mid-summer there; this is difficult and expensive, and it also happens that the sky is often hazy down
there. Another possibility is to set up a series of observing stations
around the world, spaced so that continuous observing is possible. This
is being done via the Global Oscillation Network Group (GONG).
The other possibility is to put a telescope in space, at a location that
does not have day/night cycles. This has been done on SOHO, with its
Solar Oscillations Investigation experiment. This space experiment is
in many ways the best alternative; it is also the most expensive.

Now suppose that we want to determine the flow velocity of the sun
at various depths. This information is actually crucially important for
dynamo, or magnetic field generation, models of the solar cycle. Dynamo models are theories that explain the generation of magnetic fields
inside the Sun by flow patterns of the hot interior gas, and progress
in this field has been severely hindered by lack of knowledge of the
subsurface flows.

How do observations of wave frequencies at the surface tell us anything about flows inside the Sun? Imagine that there is a layer below
the surface that is flowing from east to west at some speed. Now suppose that there is a wave of a certain frequency that travels down to that
depth before being bent back up to the surface. If we are looking at a
certain place on the surface, we will see that wave as it reemerges from
below. Moreover, we will see that wave whether it originated east of
our location or west of it. But if the wave originated west of our location, it had to flow against the subsurface stream to get to us; if it

started east of us, then it flowed with the stream. This produces a difference in the frequency of the wave when it rises to the surface, depending on whether it was slowed down or speeded up by its subsurface travel. The result is a type of frequency splitting that provides a way of measuring the flow velocities below the surface; and measurements of the different types of waves will probe different depths.

A map of the solar interior flow speeds made by this technique is shown in Figure 4.4. This figure shows the rotation rate of the Sun at the surface (far right-hand part of the figure) down into the interior of the Sun (left-hand parts of the figure). This figure shows that the interior of the Sun rotates very differently than does the surface. The Sun's surface exhibits what is called differential rotation. An ordinary globe rotates as a solid body: when you spin the globe, all of its parts take the same amount of time to complete a rotation. This is not the case for the Sun (nor for Jupiter, which also shows differential rotation). Parts of the solar surface near the equator rotate faster than parts near the pole. The right-hand portion of Figure 4.4 shows this effect (although the units need some explaining) with the equator, at 0°, rotating faster than the parts of the surface at 30° latitude, which in turn rotate faster than the parts at 60° latitude. The unit of measurement is given as nHz (nanohertz), which is to say billionths of a cycle per second. Fewer cycles per second means slower rotation, and more means faster. The rotation rate is roughly 450 nHz, meaning that the Sun completes about 450 billionths of a cycle per second. This is another way of saying that the number of seconds it will take to complete one cycle is about 1 divided by 450 billionths, or a bit over 2 million seconds. This period is roughly 26 days, which is indeed the length of time needed for the Sun's equator to complete a single rotation, as viewed from the Earth. But high latitudes are different, rotating at a rate of about 380 nHz, meaning that they require more than 30 days to complete a rotation.

Note, though, that the three curves come together and essentially coincide when we look at the interior of the Sun. When the different latitudes are all rotating at the same rate, this means that the Sun is rotating as a solid body, just as your globe does. So there is a change from

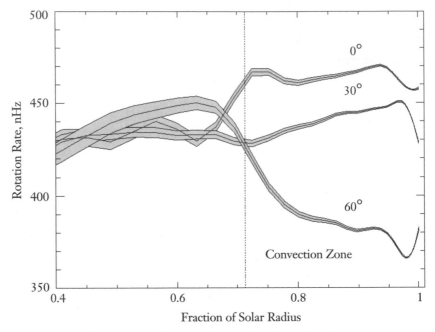

FIGURE 4.4   The pattern of flows inside the Sun, determined from helioseismology. The surface of the Sun has differential rotation, with high latitudes rotating more slowly than equatorial ones, whereas the core of the Sun rotates as a solid body. The horizontal axis shows distance from the center of the Sun as a fraction of the Sun's radius.

differential rotation to solid-body rotation deep below the visible surface, at a distance of about 72 per cent out from the center of the Sun. This change, from solid to differential rotation, was one of the predictions of the theory of the solar interior. Helioseismology therefore is able to confirm, in stunning fashion, the theoretical predictions, and also to provide details of the rotation patterns in the portion of the Sun that is believed to be responsible for producing the solar cycle.

Once these interior flow patterns are known, the next step is to construct a dynamo model to generate the Sun's magnetic fields. Such a model would be a three-dimensional simulation that would, if all goes well, reproduce the main features of the solar cycle: the 11-year oscillation in sunspot number, the reversal of the magnetic field direction in alternate 11-year cycles, the migration of sunspots from high to low solar latitudes as the cycle progresses, the start of a new cycle as the old one is dying away, and so on. This work is just beginning, so it will be at least a few years before we know whether the mystery of the Sun's activity cycle has been solved.

The SOHO experiment called the Michelson Doppler Interferometer measures velocities by observing the Doppler shift, a change in pitch or wavelength that depends on velocity. The device got its name because it uses a technique known as interferometry—based on light waves interfering with each other—developed in the nineteenth century by A. A. Michelson. Helioseismology carried out by the SOHO MDI team takes advantage of the fact that strong magnetic fields at or below the surface alter the propagation of sound waves through the subsurface regions. With good enough data, one can see magnetic flux concentrations below the surface, before they emerge (Plate XI). With this technique it is also possible to determine the presence of large sunspots on the other side of the Sun, before they rotate around to become visible to Earth-based telescopes. Both of these methods offer the possibility of extending our predictive and early-warning capabilities for outbreaks of activity.

Eclipses

The everyday sun dazzles the eye on a clear day. But about every year and a half, millions of people who are lucky enough to be in the right place on Earth see the brilliance of the Sun covered up. When even a single per cent is left visible, the sky remains blue and the event is not very spectacular. But when that last per cent disappears, the light level drops by an additional factor of 10,000, the sky turns black with pinkish color all around the horizon, the birds go home to roost, and, as observers throughout the centuries have described, "day turns to night."

A total solar eclipse is an astounding sight, one that seems to awaken primal fears. Those who see one never forget it, and often go a great distance to see another. The world of travel has advanced, and it is much easier now to fly off, as one of us did, to Chile in 1994, to India in 1995, to Mongolia in 1997, to the Caribbean island of Aruba in 1998, and to Romania in 1999, than it was for the scientists of 1936 to travel in a private railway car to Siberia, or even for Thomas Alva Edison and other scientists in 1879 to travel from the eastern United States out to Iowa.

There were 66 total solar eclipses in the twentieth century, so one every year and a half is a solid average rate. It turns out there is about the same rate of annular eclipses, when a ring—or annulus—of everyday sunlight remains visible around the disk of the Moon. And there is an additional set of partial eclipses a year. Some years have only 2 solar eclipses of any type while other years boast of as many as 5.

THE CAUSE OF ECLIPSES

Eclipses of the Sun occur when the Moon blocks our view of the Sun. Thus the Earth, Moon, and Sun lie in a straight line. Such an alignment is a form of syzygy, which is a wonderful word for an alignment of three celestial objects.

The Moon goes through its set of phases every 29 1/2 days, as it goes around the Earth while the Earth is going around the Sun. But the Moon's orbit around the Earth is tilted by about 5 degrees from the Earth's orbit around the Sun. So during most months, the Moon is either a little above or a little below the Sun when viewed from Earth at the phase we call new moon. At that time, the far side of the Moon is fully lighted by the Sun, and we see only the near side, which is then the dark side.

But several times a year, the Moon is not far in its orbit from one of the "nodes," the points where the Moon's orbit around the Earth crosses the Earth's orbit around the Sun (Figure 5.1). Then the Moon goes partially or fully in front of the Sun, and we have a solar eclipse.

The Moon's orbit around the Earth is elliptical. (The Earth's orbit around the Sun is elliptical too, but isn't as far out of round.) So the Moon is sometimes closer to the Earth than at other times. At the times when it is closer, it appears slightly larger in angle on the sky. When this occurs, and the Moon goes in front of the Sun so that their centers are close enough, we have a total solar eclipse.

At this time, the shadow of the Moon falls upon the Earth. Since the Sun is larger than the Moon, this shadow tapers gradually (Figure 5.2). As the Sun, Moon, and Earth move in their orbits, the Moon's shadow moves across the Earth. It can reach speeds of thousands of miles per hour. But the Earth rotates at the same time, and somebody on the

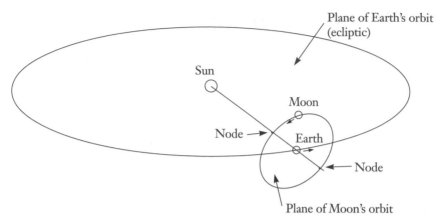

FIGURE 5.1   The relationship between Earth's orbit around the Sun and the Moon's orbit around Earth, at a time when the nodes of intersection between these two planes give solar eclipses.

Earth's surface keeps up with the Moon's shadow to some extent. The greatest speed of rotation on Earth is for an observer at the Earth's equator; after all, someone at one of the poles doesn't move at all as the Earth rotates. So the longest possible eclipse is when the Moon is closest to the Earth (making its shadow larger) and hits the Earth near its equator. Then a total solar eclipse can last as long as 7 minutes. The total solar eclipse of 1973 reached 7 minutes and 4 seconds in the Sahara Desert north of Timbuktu. Other eclipses are shorter, with eclipses of a minute or two or three being not uncommon. Still other eclipses may last only a second or so.

Thus a total solar eclipse occurs when the Moon is closest to the Earth. On the other hand, an annular solar eclipse occurs when the Moon is farthest, in its elliptical orbit, from the Earth, and the point of the Moon's conical shadow doesn't quite reach the Earth's surface. Annular eclipses can last over 10 minutes. But perhaps the most exciting kind of annular eclipse lasts only a few seconds, as did one visible from Western Australia in 1999. The mountains on the edge of the Moon may stick up enough so that an unbroken annulus is never visible, and the rapid variations in lighting can be quite dramatic.

FIGURE 5.2   A total solar eclipse, drawn to the correct scale. Within the umbra, the darkest part of the shadow, the solar photosphere is completely obscured by the Moon. Within the penumbra, part of the solar photosphere remains visible.

THE SAROS

The apparent sizes of the Sun and Moon are not the only happy coincidence that make eclipses so interesting. It also turns out that the circumstances of an eclipse—the precise geometric relationships of the Sun-Moon-Earth system—repeat 18 years and 11 1/3 days later. Thus if you have a long eclipse near the equator, you will have another long eclipse near the equator 18 years and 11 1/3 days later (or perhaps 10 1/3 or 12 1/3 days, depending on how the leap years fall). The long total solar eclipse of 30 June 1973 was followed by another long total solar eclipse on 11 July 1991, and we can confidently look forward to another wonderfully long eclipse, also with a peak duration close to 7 minutes, on 22 July 2009.

This set of coincidences is known as the saros, an interval named about 300 years ago by Edmond Halley (who mistakenly used the name adopted by the Babylonians thousands of years earlier for a different cycle). The 1/3 day in the saros has an interesting consequence: it allows the Earth to turn by 1/3 of the way. Thus the eclipses in a given saros series have to be watched from different places on Earth (Figure 5.3).

If you were out in space looking down at the Moon revolving around the Earth, you might think that it revolved in 27 1/3 days, which is how long it takes to be in the same position with respect to the stars. But the Earth's motion around the Sun means that the Moon has to rotate a bit farther to again get in line between the two, which raises the interval to 29 1/2 days. This period is known as the synodic period of the Moon. One of these periods marks a synodic month. After all, the word

Figure 5.3 The final seven total solar eclipses of a saros series, showing the characteristic saros pattern, where each eclipse path is displaced 120 degrees west of its predecessor.

"month" came from the orbit of the Moon, and is a form of "moonth."

But we can measure the Moon's progress in its orbit in other ways as well. For example, we can measure how long it takes for the Moon to go from one node in its orbit to the next, where the nodes are the places where the Moon's inclined orbit crosses the Earth's orbit. This period is known as the nodical month. Since an eclipse occurs whenever the Moon is sufficiently close to its node, these nodical months help define when eclipses occur. They are therefore also known as draconic months, after the dragon who in ancient Chinese myths devoured the Sun to make an eclipse.

As it happens, 223 synodic months is almost exactly equal to 242 nodical months: 6585.32 days versus 6585.36 days. Thus the Moon is coming back to its new moon phase after this period, and if it was near the node the last time it will be near the node this time too. Thus we have the saros of 6585.33 days, which works out to 18 years and 11 1/3 days (plus leap-year effects). The Earth has gone around the Sun almost exactly 19 times with respect to the nodes in this interval, making 19 "eclipse years" also match the saros.

In yet a further coincidence, the period that it takes the Moon to go through its elliptical orbit around the Earth, varying in distance from the Earth, matches up as well. Some 239 of these "anomalistic months" equal 6585.54 days, nearly the saros defined by the synodic and nodical months. Thus 18 years and 11 1/3 days later, not only do you have another eclipse but you also have an eclipse with the Moon approximately the same angular size in the sky. This effect means that the duration of totality at this successor eclipse is also similar. The 1973 eclipse peaked at 7 minutes and 4 seconds; the 1991 eclipse peaked at 6 minutes and 53 seconds; and the 2009 eclipse will peak at 6 minutes and 39 seconds.

The saros is not perfect, and the Moon drifts slightly south-to-north or north-to-south over the years. A whole saros series lasts up to about 1,500 years from the time when the Moon's shadow hits one of the Earth's poles to the time when the Moon's shadow drifts off the other pole.

SEEING AN ECLIPSE

Astronomers know the positions and motions of objects in the solar system so well that they can predict eclipses for thousand of years in the past and future. Over a period of a few years, they can predict the duration and timing of an eclipse for any place on Earth to an accuracy of about a second. The largest contribution to the inaccuracy comes from the shape of the Moon and the knowledge of which valleys or mountains will be on the Moon's edge as seen from Earth at the moments that totality begins or ends. The shadow of the Moon in space is a cone, and the Earth's curved surface intersects that cone in a complicated shape that is well approximated by an ellipse. Thus you can think of the Moon's shadow as an ellipse sweeping across the Earth at a speed of thousands of miles per hour. If you are within this cone, you will see a total eclipse. Since this darkest part of the cone is known as its umbra (from the Latin word for "shadow"), a total eclipse is an umbral eclipse.

If you are close enough but still off to the side of this elliptical shadow, either because it hasn't reached you yet or has passed you or because you are separated in a distance perpendicular to that of the path, the Sun will appear only partially covered by the Moon. You are then seeing a partial eclipse (or the partial phases of a total or annular eclipse), which is a penumbral eclipse, from the word for the outer part of a shadow.

The instant that the disk of the Moon first impinges on the disk of the Sun is known as first contact. It is always a relief to detect first contact, since it indicates that you are in the right place and that you have not misread the predictions. The Moon then seems to gradually cover the Sun for an hour or two. For most of that time, the eclipse is not noticeable to those on the ground, since the remaining Sun is very bright, and our visual system is built to be relatively insensitive to variations in the brightness of the Sun from clouds or haze. But as we approach second contact, when the Moon first completely covers the Sun, the situation becomes more exciting. A few minutes before second contact, it grows noticeably darker; the wind may come up and the temperature will drop. The shadows look eerie and the quality of the light changes. Indeed, we are normally used to seeing our shadows look a bit fuzzy; though we don't think about it, our shadows have narrow penumbras around the darker, umbral portions. When only a sliver of sunlight remains, the shadows appear sharper than usual and, without necessarily realizing why, we notice that they look strange. It becomes easy to understand why people have long been scared at eclipses.

With a couple of minutes remaining, you may notice shadow bands, subtle gradations of dark and light, sweep across the ground or across nearby walls. Some people even lay out a bedsheet on the ground, so as better to see the shadow bands. These shadow bands undoubtedly are caused by moving regions of air in the Earth's upper atmosphere affecting the sunlight from a narrow sliver of Sun. With the Sun fully visible, the solar image is too broad to produce clear shadow bands. Atmospheric scientists may be interested in shadow bands, but professional astronomers are more excited about waiting for the next stage.

Seconds before the Moon entirely covers the Sun, the mountains on the leading edge of the Moon stick up through the thin sliver of sunlight remaining, breaking the sunlight into bits. These bits are known as Baily's beads, after the British nineteenth-century scientist Francis Baily, though they had been seen previously by Halley and others. People around you are probably shouting for joy by the time Baily's beads are seen.

As the Moon continues to advance across the face of the Sun, the Baily's beads are extinguished. The last Baily's bead seems to glow so brightly that it looks like a diamond on a ring, producing what is known as the diamond-ring effect (Figure 5.4). (This description was first made at the 1925 eclipse that crossed New York City, with crowds gathering on street corners to mark where in the grid of streets was the dividing line between total and partial eclipse.) The diamond-ring effect commonly lasts 5 seconds or so, but if the sky is especially clear and a deep valley is aligned at the edge of the Moon, the diamond-ring effect can last for up to half a minute. If the sky is sufficiently clear, the circular band of the diamond ring may be formed by the corona.

A reddish edge to the Sun becomes visible. It got its name, "chromosphere," from its colorful appearance. It remains visible for only seconds, though its flash spectrum, described in Chapter 3, reveals many of its secrets.

Then, the diamond ring is extinguished, marking second contact. Totality begins. The crowd around you may cheer. Cameras will click away, even those of people who know better, in that they would have more fun in watching the eclipse in the sky instead of looking down at their cameras. Totality is the part of the eclipse most in use by professional astronomers, who want to study the corona that is then uniquely well visible from the ground. Although some aspects of the corona can be studied from certain high mountain tops with coronagraphs, the eclipse view is far superior. The eclipse studies from the ground are now increasingly linked with space observations of the corona, as we shall see later on.

Edmond Halley described the corona, at the 1715 eclipse he ob-

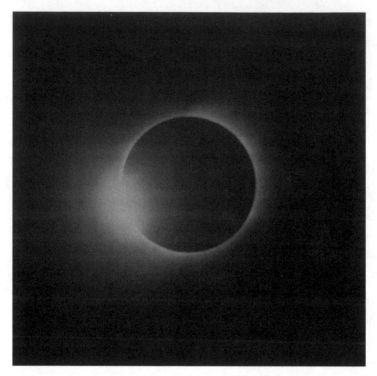

FIGURE 5.4　The diamond-ring effect, at the 1984 eclipse in Papua New Guinea.

served from England, as colored like a "pearl," and the description of the corona as pearly white remains common. The corona has a spiky appearance, with streamers coming out of the Sun's equator and thinner streamers known as polar plumes appearing nearer the poles. These beautiful structures are made of hot coronal gas held in place by the Sun's magnetic field. The shape of the corona varies over the solar activity cycle, ranging from relatively round at solar maximum, when streamers appear at all latitudes on the Sun, to relatively oblong at solar minimum, when streamers appear only near the Sun's equator.

During totality, the sky is dark in the direction of the Sun and Moon, since you are then in the Moon's shadow. The planets Venus and Mercury become visible, if they are above the horizon, and other planets and stars may be visible as well. Far from the eclipse in the sky, light is scattered into your view from regions outside the eclipse. So how dark the sky actually becomes depends on how wide the eclipse path is and on how clear the sky is. Usually, the horizon looks reddish all around, like a 360-degree sunset. As Halley wrote, "From this time the Eclipse advanced . . . when the Face and Colour of the Sky began to change from perfect serene azure blew, to a more dusky livid Colour having an eye of Purple intermixt, and grew darker and darker till the total Immersion of Sun."

After too short a time, seconds or minutes, the chromosphere and a second diamond ring appear, marking third contact. Baily's beads brighten the sky within seconds, and the exciting part of the eclipse is over. Hardly anybody bothers to watch the final partial phases, which end an hour or two later with fourth contact.

IMAGING A SOLAR ECLIPSE

Although the most exciting thing may well be to watch a total eclipse directly with your eyes, most of us succumb to the temptation to photograph it. All too many people travel to the eclipse zone and then never actually see the eclipse because they are fiddling with equipment. Still, it is fun to have your own photos.

Most cameras have lenses too short to make good pictures of the corona with an image large enough on the film. The apparent size of the Sun and Moon on your film depend on the focal length of your lens. As a rough rule, a 500-mm lens gives an image about 5 millimeters across for the Moon or for the solar photosphere on 35-mm film. So if you add an equal amount of corona on either side, such a lens fills most of the frame with the eclipse. You must have your camera on a sturdy tripod to keep it from shaking, and using a cable release so that you are not physically touching the camera is also helpful (Figure 5.5).

If you are using more conventional lenses, like 50-mm standard lenses on a 35-mm camera, then the image of the eclipse itself will be very small on the film. You may find it best to use the eclipse as a design element in the photo rather than as the central element. Put some people or objects in the foreground. Just how wide a field of view you need

FIGURE 5.5   Photo of the corona at eclipse, taken on 35-mm film with a
500-mm lens.

depends on how high the eclipse is in the sky, or on how low you can
crouch (Figure 5.6).

The corona is brightest near the solar limb and falls off rapidly with
distance outside the limb. In fact, it is over 100 times brighter near the
limb than it is only 1 solar radius out. (Solar astronomers often talk in
solar radii, merely meaning half the angular diameter of the Sun in the
sky.) We would like to record the corona out to at least several radii and
film doesn't easily cover that much range in brightness, so photos will
capture only a limited part of the corona at the proper exposure and
will be overexposed or underexposed for other parts.

The American astronomer Gordon Newkirk built a camera in the

FIGURE 5.6   Photo of the corona at eclipse, taken on 35-mm film with a wide-angle lens.

1960s for capturing more of the brightness range on a single piece of film. He and his colleagues, mostly from the High Altitude Observatory in Boulder, Colorado, using a previously developed technique, made a filter that had absorbing and reflecting material deposited on it. The filter was densest in a ring the size of the Moon, and became less and less dense outward in the radial direction. It was called a radially graded filter, or, for short, a radial filter. This filter was used in a device, commonly called a Newkirk camera, to make images of the corona at a series of eclipses (Figure 5.7). The French astronomer Serge Koutchmy has also made a series of radial-filter images, as has the American Jonathan Kern.

The standard Newkirk camera did more than merely take these pretty pictures; it also made observations of the polarization of the corona, and the observations were carefully calibrated so that scientists could study the density of electrons in the corona. By the 1998 eclipse,

FIGURE 5.7   Ten solar eclipse images made with the Newkirk camera by the staff of the High Altitude Observatory of the U.S. National Center for Atmospheric Research, which is operated by the University Corporation for Atmospheric Research and sponsored by the National Science Foundation; the 1980 and 1991 expeditions were conducted jointly with Rhodes College. The last image uses a numerical processing method to replace the radially graded optical filter. Note how the overall shape of the corona varies with the solar activity cycle—rounder at solar maximum and more elongated at solar minimum.

FIGURE 5.7A
12 November 1966

FIGURE 5.7B
7 March 1970

FIGURE 5.7C
30 June 1973

FIGURE 5.7D
16 February 1980

FIGURE 5.7E
31 July 1981

FIGURE 5.7F
11 June 1983

FIGURE 5.7G
18 March 1988

FIGURE 5.7H
11 July 1991

FIGURE 5.7I

3 November 1994

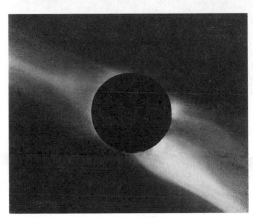

FIGURE 5.7J

26 February 1998

the original Newkirk camera had been retired, and a camera using an electronic detector was used instead.

As techniques were developed for scanning film, it became possible to make composite photographs out of several individual ones, each exposed to show a different brightness region of the corona. The longer the exposure, the farther out in the corona you see. In the 1990s, these techniques were perfected so much that composite images showed well the coronal structure of a wide range of solar radii outside the limb (Figure 5.8).

Through that decade, film became replaced by electronic detectors in much of astronomy, and eclipse photography was no exception. Most of the detectors are charge-coupled devices (CCDs). Even the

FIGURE 5.8   The 1999 total solar eclipse, a composite of CCD images of various exposure time. Since the eclipse occurred very close to the maximum of the solar activity cycle, streamers were visible at all solar latitudes instead of being restricted to low latitudes.

camcorder you see advertised in your Sunday newspaper has a CCD in it, though the ones professional astronomers use are more carefully made as well as being more precisely calibrated and being more controllable in exposure time and other functions.

Many people now make videos of eclipses, using their CCD camcorders to make the tapes. Ordinary VHS camcorders give satisfactory images, though Hi-8 and S-VHS camcorders have higher quality, and digital camcorders are giving higher quality still. These standard consumer cameras usually have zoom lenses so powerful that the corona can be made to sufficiently fill the frame. A hint is to focus the camera manually in advance, since the automatic focus feature would otherwise "hunt" back and forth, losing valuable time during the all-too-brief eclipse.

Note that only during totality, from diamond ring to diamond ring, can you look at the Sun directly or can the camera image the Sun directly. Before second contact or after third contact—during the partial phases—you need to use special solar filters to reduce the intensity of the solar photosphere to safe levels for your eye or for your camera.

So the routine for taking photographs of an eclipse is to have the filters in front of your camera lens until the diamond ring appears, and then to take it off for the interval of totality. It must then go on when the second diamond ring appears. Just how well you handle that timing affects how good your eclipse video will be.

EYE SAFETY AT ECLIPSES

All too many people think that eclipses are hazardous in some special way, perhaps because some special dangerous rays come out of the corona. That is not the case. The eclipse is a time when there is less light coming out of the Sun, rather than more of anything. The corona is always there, but you can't normally see it because the photosphere is too bright.

The everyday solar surface—the photosphere—is too bright to look at safely, and as long as any of the photosphere remains visible, its surface brightness can cause damage to the eyes of people who stare at it or who look at it even briefly directly through an optical device such as an unsatisfactorily filtered telescope or binocular. The eye normally alerts you to excessive total brightness, but it is fooled at an eclipse when only a tiny piece of very bright photosphere remains, because your eye does not then sense the danger.

You can buy a special solar filter that reflects or absorbs all but one part in 100,000 of the incoming sunlight. The ones made out of Mylar (plastic) with a coating of aluminum on them can be found for about a dollar. Still, we should use such filters only for a glance of a second or two every few minutes during the partial phases. There is no point in staring for longer at a partially eclipsed sun: it isn't fundamentally interesting and it is best to wait for totality, when you put your filters down and glory in the direct view of the solar corona.

The solar corona is about as bright as the full moon, and equally safe to look at. So for the period of totality, and only for that period, you can and should look at it directly, without filters.

During the partial phases, you don't even have to use a solar filter to keep track of how the eclipse is going. You can make a simple pinhole camera instead. To do so, merely punch a hole 1/4-inch or so in a piece of paper or cardboard (or, a little better, in a piece of aluminum foil). Then hold that piece with the hole in it up in the air, and allow it to project an image of the Sun onto a piece of paper or cardboard held about two feet down toward the ground. You will see a crescent sun on that piece of paper, and since you are looking down at it, with the Sun at your back, it is completely safe to look at. Of course, you don't look up at the Sun through the hole.

HISTORICAL ECLIPSES AND THE ROTATION OF THE EARTH

Although astronomers can predict the positions of the Sun, Earth, and Moon thousands of years into the past or future, one confounding factor in predicting eclipses is that the rotation of the Earth is not regular enough to allow the exact location of the Moon's shadow on the Earth to be accurately known over periods of many millennia.

F. R. Stephenson of the University of Durham in England and his colleagues have surveyed historical reports of eclipses in various parts of the globe for thousands of years. From their work, they have found how the Earth's rotation has varied over time, gradually slowing down. Their results have importance for geology, since they reveal processes under the Earth's surface that affect the slowdown. For example, after the era of glaciers, uplift of the ground may have changed the Earth enough to affect its rotation in this way. On the basis of observations over the past 2,500 years it has been found that the dominant effect is a slowing of the rotation due to tidal friction, increasing the length of the day at an average rate of about 2.3 milliseconds per century. There is also a variation, perhaps due to post-glacial uplift, toward a shorter length of day by 0.6 milliseconds per century, giving a net effect of about +1.7 milliseconds per century. Thus over the last 2,500 years, the day has increased in length by about 42 milliseconds, or about 1/25 second.

Accurate prediction of both the time and the location of solar eclipses is a relatively recent phenomenon. The story is often related in books on eclipses how a total solar eclipse on 28 May 585 B.C.E., predicted by the Greek astronomer Thales of Miletus, stopped a war between the Medes and the Lydians. The story was related by the historian Herodotus (ca. 484–425 B.C.E.), who wrote about 150 years after the event. Herodotus wrote, "When the Lydians and the Medes saw the day turned to night, they ceased from fighting, and both were the more zealous to make peace." But Stephenson's evaluation is that the account is of "doubtful reliability," and that the methods of predicting eclipses were not far enough advanced in 585 B.C.E. for the story to be true. Stephenson concludes that "the assertions by Herodotus and Pliny probably owe their origin to the various legends which accumulated around the personality of Thales."

Recent and Future Solar Eclipses

The locations from which people see total eclipses may seem randomly distributed (Figure 5.9). If you stand in one place, it may take centuries for a total eclipse to be visible overhead. Fortunately, it is relatively easy to travel these days, to take advantage of eclipses much more often than that.

On 21 June 2001 a relatively long total eclipse crossed southern Africa and Madagascar before heading out into the Indian Ocean. Totality lasted almost 5 minutes over the Atlantic Ocean and over 3 minutes at most of the ground-based observing locations. The weather was

FIGURE 5.9   Total eclipse paths traced on the Earth's surface for the years
1996–2020.

exceptionally clear. Another total eclipse crossed southern Africa 18
months later, on 4 December 2002. This eclipse lasted only 2 minutes
at maximum. Near the end of the path, near sunset, it reached southern
Australia, west of Adelaide.

The total eclipse of 23 November 2003 will be visible only from
Antarctica and the ocean near it. Historically, such an eclipse might not
have been viewed by anybody, but now eclipse buffs are so excited
by eclipses that some expeditions by ship or plane are already being
organized.

Table 5.1 lists the paths of total solar eclipses for the decade 2001–
2010. The next eclipse that crosses United States territory won't be
until 2017. Continental Canada won't have a total eclipse until 2024.

Annular eclipses aren't as spectacular as total eclipses, but can be fun
to watch. Filters stay on for the whole eclipse, including annularity
(Figure 5.10).

FIGURE 5.10 Annular eclipse paths traced on the Earth's surface, for the years 1996–2020.

THE SCIENTIFIC VALUE OF ECLIPSES

People often ask whether, in this age of space observations, there is still scientific value to observing eclipses. The answer is that surely, yes, there are still many things to be learned from solar eclipse expeditions on Earth that cannot yet be found from space observations. And there are uses for making coordinated observations from the ground and from space.

Similarly, for nonsolar astronomy, the question is often asked why ground-based telescopes are necessary now that the Hubble Space Telescope is in orbit. The answer to that question is that Hubble is only one small telescope, and that there is lots to do separately from Hubble and also lots to do in following up problems suggested by Hubble observations.

The eclipse of 1999, for example, was the first eclipse when the x-ray

Table 5.1 Total solar eclipses, 2001–2010.

Date	Maximum totality (min:sec)	Path width (km)	Locations
21 June 2001	4:57	200	Southern Africa, Madagascar
4 December 2002	2:04	87	Southern Africa, Australia
23 November 2003	1:57	496	Antarctica
8 April 2005	0:42	27	Total only at maximum: Pacific Ocean Annular: Panama, Colombia, Venezuela
29 March 2006	4:07	184	West and North Africa, Turkey, Russia
1 August 2008	2:27	237	Northern Canada Islands, Greenland, Siberia, Mongolia, China
22 July 2009	6:39	259	India, Bangladesh, China; peak over Pacific Ocean
11 July 2010	5:20	262	South Pacific

telescope Yohkoh, the ultraviolet and visible-light telescope on SOHO, and the ultraviolet and visible-light telescopes on TRACE were all aloft; and imaging the solar corona from ground-based eclipse sites fills in some gaps in the coverage from space. For example, the corona-graphs on SOHO—and most other coronagraphs as well—have to occult (hide) not only the photosphere but also a region around the photosphere about half a solar radius in extent, in order to limit the amount of scattered photospheric light that contaminates the coronal image. It is exactly this inner corona region—missing from the space-based data—that is well imaged by ground-based eclipse observations, because the Moon provides an ideally placed occulting disk having only minimal scattering and diffraction problems (Figure 5.11). Combining ground-based observations at eclipses with the space-based corona-graphs also has the potential of improving knowledge of the character-istics of the light scattered around in the space telescope, improving the interpretation of data taken with it.

The TRACE spacecraft makes exquisitely detailed observations of loops of hot coronal gas held in place by the Sun's magnetic field. But

FIGURE 5.11   A composite image with a SOHO EIT image at the center, a com-
pound eclipse image surrounding it, and an image taken with one of the LASCO
coronagraphs on SOHO on the outside. The eclipse image fills in the gap in
SOHO's coverage of the corona, but only for the limited time of the eclipse.

the instruments typically observe at a cadence of at least a few seconds
between exposures. So an experiment that images the Sun over and
over again at 10 times per second, such as one of the eclipse experiments run by one of us, fills a part of observational space that is not
otherwise covered.

The questions to study at a ground-based eclipse expedition must be
carefully chosen to complement or to be separate from observations
made from telescopes in space. But the ability to use new equipment,
perhaps to study newly developed ideas, and to do so at a relatively low
cost compared with that of space observations, leaves ground-based
eclipse studies a role to play for the forseeable future.

Of course, most people seeing solar eclipses are not professional astronomers. It is unusual to have a total eclipse pass near you, but most of the people seeing any individual eclipse are just those people who are near home. The 1999 eclipse was notable in that it crossed Europe in a path that was close to the homes of tens of millions of people. Increasing numbers of tourists, many with no special interest in or background in astronomy, are traveling in groups to see eclipses. Eclipse cruises have become big business, with many ships carrying tourists to the Caribbean for the eclipse of 1998 or to the Black Sea for the eclipse of 1999. Many land tours take people for 2 weeks or so to a variety of sights, making sure only to be in totality at the proper time on eclipse day. The eclipse itself—not only the corona in the sky but also the darkening of the daytime sky, the change in wind and weather, and other phenomena that are observed—always leaves the traveler feeling rewarded.

THE CORONA

With the advent of photography in the nineteenth century, the corona became a familiar sight, especially when imaging techniques improved, as described above. But without photography, or some other means of recording the image, very few people would ever see a total solar eclipse, because the ribbon of area in full umbral shadow is so narrow. There is ample historical evidence that total eclipses of the Sun were known thousands of years ago, but it is unclear whether there is any historical evidence prior to the seventeenth century that anyone knew that the Sun has a corona. An example of how ambiguous the historical record can be is an alleged report of a Babylonian eclipse in 1063 B.C.E.: "On the twenty-sixth day of the month of Sivan in the seventh year the day was turned to night, and fire in the midst of the heaven." This is the sort of record that would, a few years ago, have been taken to indicate not only a total eclipse (day turned to night) but a sighting of the corona as well (fire in the heaven). But today historians are more cautious. In analyzing this and other reports, the historian Roger Newton points out: "If the passage refers to the eclipse of −1062, both the month and the day of the month are given wrong. The passage does not mention the Sun directly and it does not use the standard Babylonian term for an eclipse of the Sun. There are plausible explanations for all these discrepancies, but the combination of so many of them in one record does not inspire confidence."

One of the first clear mentions of something that sounds very much like the corona was made in the seventeenth century by Kepler, who specifically asked: "Why does the Moon remain partly visible when the Sun is behind it during total eclipse?" Kepler, however, seemed to think that the corona is caused by scattering of sunlight in a lunar atmosphere: "the whole Moon is surrounded by some sort of aery essence, which reflects rays from all parts."

This question of whether the glow around the Moon does or does not belong to the Sun was not settled until more than two centuries later. Since it had already become clear by then that the Moon does not have an atmosphere, the lunar explanation had been ruled out. Could the glow around the Moon then be caused by the Earth's atmosphere? Photographs of the corona were taken during the eclipse of 1860 from widely separated sites. These showed that the features seen at both locations were the same, providing good evidence that the phenomenon was connected to the Sun. Then in 1869 a strong emission line coming from the corona was found in the green part of the spectrum, which was taken as conclusive evidence that the corona is solar—such an emission line could only come from a hot gas. But this discovery, in turn, raised a problem that took 50 years to resolve: why is the corona emitting this spectral line?

Observations of the 1868 eclipse led to the discovery of a bright yellow emission line in the spectrum of the chromosphere, which is normally not observable except during a few seconds just before and just following totality. What happened next is nicely described by C. A. Young in the 1895 edition of his book *The Sun:*

> The famous D_3 line was first seen in 1868, when the spectroscope was for the first time directed upon a solar eclipse. Most observers supposed it to be the D line of sodium, but Janssen noted its non-

coincidence; and very soon, when Lockyer and Frankland took up the study of the chromosphere spectrum, they found that the line could not be ascribed to hydrogen or to any then known terrestrial element. As a matter of convenient reference Frankland proposed for the unknown substance the provisional name of "helium" . . . Naturally there has been much earnest searching after the hypothetical element, but until very recently wholly without success . . . The matter remained a mystery until April, 1895, when Dr. Ramsey, who was Lord Rayleigh's chemical collaborator in the discovery of argon, in examining the gas liberated by heating a specimen of Norwegian cleveite, found in its spectrum the D_3 line, conspicuous and indubitable . . . Cleveite is a species of uraninite or pitch blende, and it soon appeared that helium could be obtained from nearly all the uranium minerals.

As we now know, the connection between uranium and helium is that radioactive decay of uranium involves what were at that time called alpha particles, which are helium nuclei. These nuclei pick up electrons to become atoms of helium, which can become trapped in uranium-rich rocks, to be released when the rocks are heated.

Meanwhile, in 1869, a year after the discovery of the helium line on the Sun, Young and Williams Harkness, working independently, found a strong green coronal emission line that could not be identified (Figure 5.12). As had been done the year before, a new element, named coronium, was proposed. To quote Young again, writing in 1895: "the name 'coronium' has been provisionally assigned to it, and the recent probable identification of 'helium' in terrestrial minerals gives strong reason to hope that before very long we may find coronium also."

This new element had a most unusual property: it had to be much lighter than hydrogen, the lightest known element. The reason for this is that the corona—the Sun's atmosphere—is seen to extend very far out from the solar surface. The height of an atmosphere is determined by the balance between the downward pull of gravity and the upward mobility of the atmosphere's gas particles from their temperature. For a given temperature, lighter particles move more quickly than heavy particles. In essence, temperature is a certain amount of energy of mo-

FIGURE 5.12 Ground-based images of coronal loops taken in the "coronium" green line.

tion per particle, and light particles need to move faster than heavy ones to have the same amount of energy. Like a ball tossed in the air, the particles move upward to some height—determined by their initial upward velocity and by the downward pull of gravity—and then they fall back down. Averaging over all of the particles in the atmosphere, we find that the density of the gas falls off as we move upward, gradually becoming less and less.

So there is a quantity called the scale height, which contains the ratio between the temperature of the gas and the mass of the particles. This ratio tells us how far upward the atmosphere will extend. For the corona, the scale height for any ordinary element, even hydrogen, at a photospheric temperature is so small that only a thin skin sitting just above the solar surface would be visible. Since the corona extends upward to great distances, while its spectrum seems to show a photospheric temperature, we are forced to conclude that coronium must be an extremely light element, much lighter than hydrogen.

This conclusion posed a tremendous problem, especially after Dmitri Mendeleev's work on the periodic table of the elements. In 1871, Mendeleev showed that the elements form a systematic arrangement, and it soon became clear that there was no room for coronium in the pattern. The choice was then either to question the periodic table or to get rid of coronium. But then what was causing the extended coronal emission?

The fortunes of coronium were further damaged by Ernest Rutherford's work in 1911, showing that atoms consist of a small, dense nucleus surrounded by lightweight electrons. Simple counting implies that hydrogen, with only a single nucleon, is as light as an element can be. So now there was no room, even in principle, for coronium.

Many explanations for the mysterious coronal emission lines (several were known by then) were proposed in the early decades of the twentieth century, but it was not until 1939 that the answer was finally found. It came through the convergence of many separate avenues of research, ranging from laboratory experiments on the spectra of spark discharges to astronomical studies of nebulas (for which the element "nebulium" had been hypothesized). As often happens, once the answer is known, one feels that it should have been obvious all along: the corona is extremely hot.

The reason that the coronal emission lines were so hard to identify is that they arise from gas in a state that is not normally encountered, and therefore one that investigators had not experienced. As a gas is heated, the violent collisions between atoms and free electrons in the gas become strong enough to knock bound electrons up into higher energy levels of the atom (a process called excitation), and sometimes to knock them out of the atom completely (ionization). These processes had been well studied throughout the nineteenth century, but for relatively low-energy versions and mainly for ionization of the neutral atom, the atom supplied initially with all of its electrons.

Studies were gradually progressing toward more exotic conditions, in which more than one electron is removed from an atom, and the

PLATE I An assortment of planetary nebulas, as imaged by NASA's Hubble Space Telescope.

Intensity

PLATE II   Intensity, corrected intensity (modulation), and velocity images from the Global Oscillation Network Group (GONG).

Modulation

Velocity

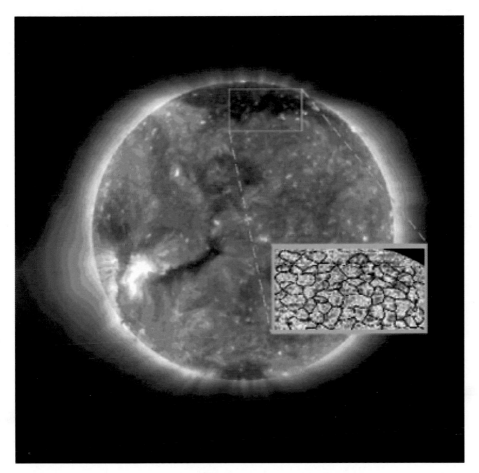

PLATE III   The solar wind in coronal holes originates in the lanes between
supergranules.

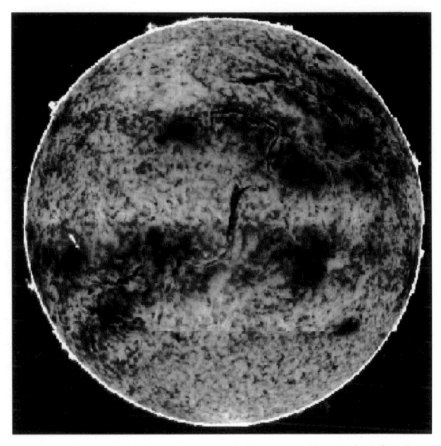

PLATE IV   Image of the Sun in the neutral helium spectral line in the infrared at a wavelength of 1,083 nanometers (billionths of a meter), 4 March 2000 (1,083 nanometers is a wavelength just over 1 micrometer, that is, one millionth of a meter and one thousandth of a millimeter).

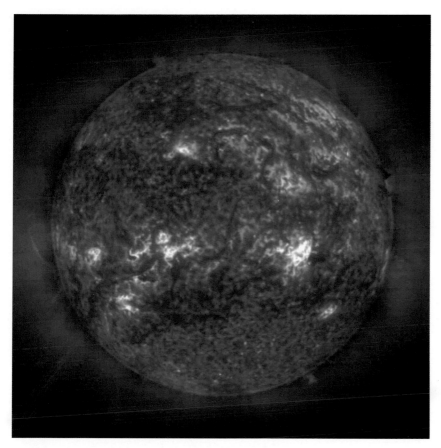

PLATE V   Image of the Sun in the He II Lyman-alpha line at 30.4 nanometers (billionths of a meter), taken by the Extreme-ultraviolet Imaging Telescope (EIT) on the Solar and Heliosphere Observatory (SOHO), 4 March 2000.

PLATE VI   Image taken in the Lyman-alpha line of hydrogen, 19 May 1998, from the Transition Region and Coronal Explorer (TRACE) satellite.

PLATE VII   A visible-light composite image of the 11 August 1999 eclipse with an inserted central image from the SOHO/EIT, which directly images the extreme ultraviolet (EUV) emission from the hot corona.

PLATE VIII   A view of the 1,000,000 K corona, as seen by the TRACE satellite, 19 May 1998.

PLATE IX A large solar flare, viewed in ultraviolet light by the TRACE satellite.

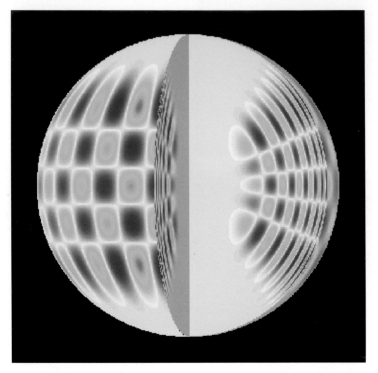

PLATE XA   A mathematical model representing one of the many
modes of solar oscillations.

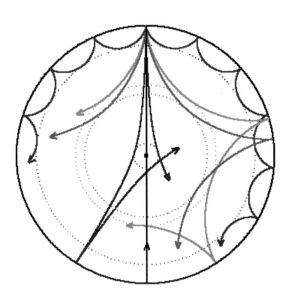

PLATE XB
Waves traveling
into the Sun are
bent back up toward
the surface, where
they can be studied
to reveal properties
of the interior.

PLATE XI   Sunspot data from the high-resolution mode of the Michelson Doppler
Interferometer (MDI) experiment on the Solar and Heliospheric Observatory (SOHO),
18 June 1998. Sunspots show where magnetic flux is concentrated.

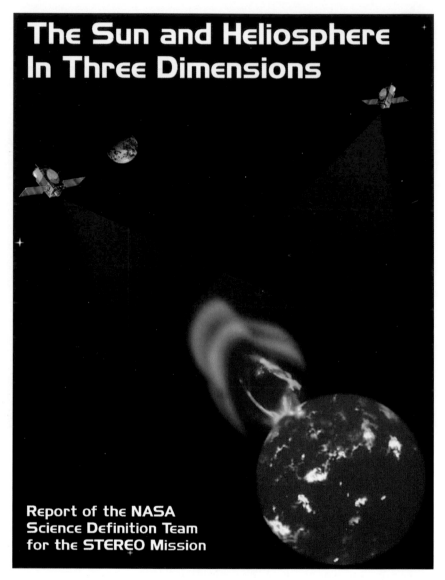

The Sun and Heliosphere
In Three Dimensions

Report of the NASA
Science Definition Team
for the STEREO Mission

PLATE XII Cover art from the STEREO Working Group report, showing a conception of the planned STEREO mission viewing an Earth-directed coronal mass ejection from the Sun.

PLATE XIII Top: The planned Solar-B satellite, including the Solar Optical Telescope (SOT), the Focal Plane Package (FPP), the EUV Imaging Spectromcter (EIS), and the X-ray Telescope (XRT). It is powered with Solar Array Panels (SAP). Bottom: Close-up view of the XRT.

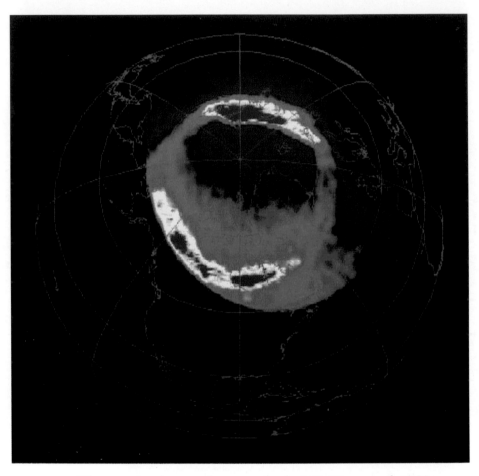

PLATE XIV The auroral oval, as seen from space by the POLAR spacecraft, 27 May 1996.

PLATE XV A solar flare, observed by the Transition Region and Coronal Explorer
(TRACE) satellite, 14 July 1998.

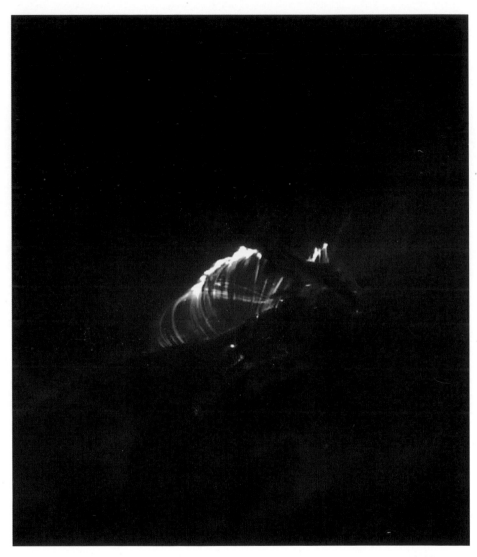

PLATE XVI The eruption of a filament (dark in this image) causes magnetic field reconnection and energy release in the corona; observed by the TRACE satellite, 30 September 1998.

prevailing conditions (such as high temperature and low density) are such that the atom remains in a multiply ionized state for a long time. This turned out to be the path toward explaining the coronal spectrum. The corona contains emission lines from such unfamiliar states as, for example, 9 times ionized iron, and 14 times ionized calcium. These are not at all obvious choices for explaining the observed spectral lines, and even once they were suggested, the calculations needed to predict the wavelength of, say, the coronal green line, are difficult.

On the basis of a long series of laboratory experiments by the Swedish spectroscopist Bengt Edlén and by Ira Bowen at Caltech, the "nebulium" emission lines were identified as coming from highly ionized states, such as 5-times ionized iron. This work gave Walter Grotrian, in Potsdam, the clue to the coronium lines. He suggested that 9 and 10 times ionized iron would produce spectra that included rare atomic transitions at visible wavelengths. When Grotrian's brief note on the subject was published in 1939, Edlén looked over his laboratory work and noticed that his spark discharge spectra supported this view. Very quickly he determined that the famous green line could be explained by one of these rare optical transitions in 13 times ionized iron. The rest, as they say, is history.

Once the breakthrough was made, the corona was once again a physically possible object, and the problem became one of explaining not what it is, but how it gets that way: why is the corona so hot? This is still one of the great outstanding puzzles in solar research today.

Early investigators did not immediately realize how hot the corona is because there are at least four different types of light coming to us, and only one of them is actual emission from the hot coronal gas. The others are due to scattering or reflection of the extremely bright photospheric light, with their brightnesses varying with the distance above the Moon's disk (Figure 5.13). They are:

K-corona   From the German word "Kontinuierlich," or continuous, because it emits light over a broad spectrum, this is the brightest component of the corona in the visible. It arises out of photospheric light that is scattered by the

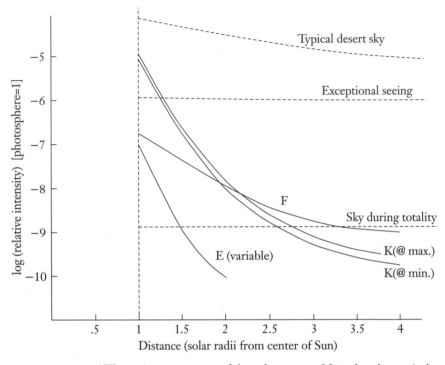

FIGURE 5.13   The main components of the solar corona. Note that the vertical scale is logarithmic (powers of 10) and is relative to the brightness of the solar photosphere. The corona is generally below −6, or one millionth, on this scale.

electrons in the corona. The spectrum appears continuous, even though the photosphere has the strong absorption lines found by Fraunhofer, because the electrons are moving so quickly in random directions in the corona that the lines are smeared out, becoming so broad that they are no longer detectable as lines.

F-corona   So named because it shows the dark absorption lines found by Fraunhofer, this component is due to scattering of the photospheric light by grains of dust in the ecliptic plane. The solar system is filled with such dust, which also causes the faint glow at night known as the zodiacal light. Because the dust particles are moving relatively slowly, the light reflected from them is not very much shifted in wavelength relative to the incoming light, and the dark lines remain discernible in the scattered light. The F-corona is not coronal light, although it happens to be about the same brightness, and it also falls off in intensity as one moves away from the disk of the Sun.

E-corona   The emission-line (hence "E-") corona is the only component in this list that is actually due to emission of light from the hot coronal gas. It is due to the excitation of electrons in atoms of the corona, followed by atomic transitions of those electrons back down to lower energy levels, with the consequent emission of light. The emission-line component of the corona is hard to see in broad wavelength recordings of the coronal light—such as photographic plates—but it can be detected clearly if a narrow slice of the spectrum is selected for viewing. Within this small range of wavelengths—which might be anywhere from the infrared through the visible and ultraviolet into the soft x-ray region of the spectrum— the E-corona is relatively bright, and can be recorded for scientific study.

T-corona   This newly recognized component of the corona is now detectable because of improvements in infrared viewing techniques. The interplanetary dust that causes the F-corona and the zodiacal light is warmed by the Sun and therefore emits radiation due to its temperature. This thermal (hence "T-") emission adds to the continuum of the spectrum in the infrared. Since it is in the infrared rather than the visible, it does not show on Figure 5.13.

Only the E-corona would show obvious evidence that the corona is hot, and it is quite faint except for the "coronium" emission lines, which were seen in the nineteenth century, but which could not be identified at that time. The F- and K-coronas were dominant in photographic images and in low-resolution spectrographs, and they are due

FIGURE 5.14   The spectrum of the coronal plasma at 3 million degrees. The light emitted is about 100 times shorter in wavelength than visible light. The horizontal scale is in nanometers; multiply by 10 to change to angstroms.

to photospheric light. So it is understandable that the corona looked photospheric, except for the puzzling emission lines.

Although the first true coronal emission that was detected was visible light, this was a historical accident rather than something fundamental about the corona itself. In reality, because of the corona's high temperature, the bulk of the radiation it emits occurs at much shorter wavelengths than those of visible light, in the regions called the extreme ultraviolet and also shorter wavelengths known as soft x-rays ("soft" because they have much lower energy and much less penetrating power than, for instance, medical x-rays).

Figure 5.14 shows the spectrum of the corona at x-ray wavelengths, based on a complex computer code that takes into account dozens of different constituent elements in the corona, dozens of atomic energy levels in each element, and thousands of interactions among all of them, all as a function of temperature. Note that the scale is in nanometers, and that the typical wavelength emitted is about 2 nanometers. For visible light, the typical wavelength would be about 500 nanometers. Different parts of the corona are at different temperatures, ranging from 1,000,000 to 10,000,000 degrees, and the spectrum varies with temperature. This spectrum is for a piece of the hot corona in active regions, at a temperature of 3,000,000 K (about 5,000,000°F).

Such calculations show us what the emission is from the hot coronal gas, but not what it *does*. More is needed if we want to know how the gas is heated and why it behaves so dynamically. We need a way to produce an image of the x-ray emission. A telescope that is specially designed to focus x-rays is needed for these wavelengths. Three such instruments, the existing TRACE satellite, and the X-ray Telescope (XRT) and EUV Imaging Spectrometer (EIS) on the planned Solar-B satellite, will be discussed in the next chapter.

Space Missions

Putting observing instruments into space is difficult, expensive, nerve-wracking, and often frustrating. Launch opportunities are limited, the risk of failure is high, and it usually requires many years for a project to reach completion. Why, then, would anyone want to do science this way?

The problem is that the Earth has an atmosphere. This is good news for humans, of course. Life as we know it would not be possible without an oxygen-rich atmosphere, which also acts as a shield against harmful solar radiation and energetic particles known as cosmic rays (which are also deflected up toward the poles by the Earth's magnetic field—another type of shield). Our atmosphere prevents most meteors from hitting the ground—except for an occasional larger one—and it provides a greenhouse warming that brings global temperatures up to habitable levels, as we will discuss in the next chapter.

Most of these benefits to life are bad news for astronomers. From ground level we do not have direct access to cosmic rays, so it is difficult to study their origin; we do not have direct access to the bits of dust and rock floating through the solar system, to help determine the history of the formation of the planets; we cannot, from the ground, directly measure the solar energetic particles to study the origin and effects of solar storms. Merely trying to look at astronomical objects is difficult: the atmosphere is murky and turbulent, it is full of dust and other obscuring materials, and scattered light from human habitation prohibits observations from most populated areas. The best views of these objects are therefore obtained by putting the telescopes above the atmosphere.

Our atmosphere is also quite opaque to most wavelengths of light, as we showed in Figure 4.1. It is only the very narrow slice of wavelengths in what we call the visible that penetrates the atmosphere fairly well. But astronomical objects emit energy at wavelengths from the radio through the infrared, visible, ultraviolet, x-rays, and even gamma rays. With our extremely limited sensory equipment and sitting at the bottom of our murky ocean of atmosphere we are literally blind to most of what is happening out there.

Another major reason to go above our atmosphere is to decrease the background noise level. An example may help to explain why this matters. If you are at a rock concert, with the performers using 10,000-watt amplifiers, you will not be able to hear a person next to you speaking or even shouting in your ear. On the other hand, if you are in the reading room of a library, someone whispering across the room may seem loud enough to be disturbing. The important quantity in these comparisons is the signal-to-noise ratio, the ratio of the intensity of the desired information (signal) to the undesired information (noise). If the noise level is reduced, we can more easily detect a weak signal.

Our atmosphere is never completely dark, even at night. Sunlight and city lights scatter in air, so that there is always some amount of light coming down to us from the sky above. This can be seen in a long-exposure photograph, showing that there is enough illumination at night to take a picture, if the shutter is left open long enough. But astronomers looking at faint objects do need to take long exposures, sometimes several hours long, so that the sky brightness becomes a factor limiting the faintness of the targets that can be seen.

Solar astronomers trying to record the corona encounter this problem. The corona, even near the Sun, is faint, and special instruments

FIGURE 6.1   Combined LASCO (C1, C2, and C3) and EIT images showing the
extended corona on 17 April 1998.

are needed to detect it, as we described in the last chapter. The corona
quickly becomes even fainter as we look farther away from the Sun, and
the background sky is very bright since, by definition, we are observing
in the daytime. If we want to study the corona as it leaves the Sun, or as
hot plasma is ejected from the Sun toward the Earth, we need to have a
lower sky background, and this is only obtained in space. A nice example of the advantages gained by placing a coronagraph in space is provided by the view of the extended corona recorded by the LASCO and
EIT instruments on SOHO, shown in Figure 6.1.

There are also reasons to carry out observations in the infrared, at
wavelengths longer than the visible. First, a large fraction of the energy
from the Sun and stars is emitted below the visible, and there are also

features on the Sun and other astronomical objects (such as star formation regions) that are best seen in the infrared. But our atmosphere is at
a finite temperature, which means that it is a source of infrared radiation itself. If our eyes could see these wavelengths, the night sky would
not look dark at all, but would be glowing like the burner on your
stovetop. So again, an entire slice of the spectrum becomes available by
putting your telescope above the atmosphere, or at least very high up
on a mountaintop or with a large balloon in order to be above as much
of the atmosphere as possible. Note that the telescope mirror itself
must be cooled, or it glows in the infrared as well!

THE VIEW FROM SPACE

Solar studies from space have been concentrated mainly in two areas:
direct measurements of particles coming from the Sun, and observations of the invisible wavelengths of light in the ultraviolet (UV), soft x-
ray, and the very faint visible parts of the spectrum that are undetect-
able from the ground. Historically, the particle measurements arose
out of the many years that had been spent studying cosmic rays, and the
invisible wavelength measurements came about mainly because of the
Sun's influence on radio communications.

The radio studies commenced with the early "wireless" transmis-
sions of Guglielmo Marconi in 1901. Signals were sent succesfully from
England to Canada, even though the curvature of the Earth prevents a
direct line-of-sight view between such widely separated locations. (It is
not clear why he tried this in the first place, given that it should not
have worked.) The explanation for this long-range transmission, arrived at independently by an American engineer, Arthur Kennely, and
the British physicist Oliver Heaviside, is that a layer of the upper atmosphere is highly ionized—it contains a large number of mobile electrons, giving it a high electrical conductivity—so that the atmosphere
acts as a conducting waveguide that reflects the radio signal back down
toward the ground. (Heaviside is immortalized in the Heaviside layer
in the Broadway musical *Cats*, which is based on poems by T. S. Eliot.)
This layer was later given the name ionosphere. The obvious next

question is, what is causing this ionization of the upper atmosphere? The answer came largely through the work of Edward Hulbert, head of the Heat and Light Division of the newly formed Naval Research Laboratory (NRL) in Washington, D.C. Hulbert established that there were day/night and seasonal variations in the ionosphere, and coupled this with the disruptions in radio communications following solar flares, to suggest that the Sun was the source of the ionosphere. In particular, he suggested in 1938 that x-rays from the Sun were the main contributor to the ionization, the basic idea being that the incoming x-rays are absorbed by the first traces of gas high in the atmosphere and transfer their energy to these atoms, resulting in an electron being "kicked out" of the atom every time an x-ray is absorbed.

Testing this theory required direct, or *in situ*, measurements. Unfortunately, the ionosphere is formed at altitudes between 80 and 225 kilometers (50 to 140 miles), far higher than could be reached by any available methods at the time, such as aircraft or balloons. But in 1945 the U.S., and later the Russian, armies took over control of the German V-2 missile factory near Peenemünde, and shipped trainloads of rocket parts (and rocket scientists) back to their respective home countries. In the United States, a rocket test site was set up near White Sands, New Mexico, and an offer was made to NRL (which had presciently just created a Rocket Sonde Research branch) to use the V-2s for scientific studies. The early flights brought out difficulties, such as recovering instruments that had been turned into mangled scraps of metal at the bottom of a large impact crater when the rocket returned to Earth at many times the speed of sound.

By 1952, Herb Friedman of NRL had shown that solar x-rays are indeed responsible for the formation of the so-called E-region in the upper atmosphere. Friedman also branched out into night-time rocketry and helped establish the field of x-ray astronomy, although it was a group in Cambridge, Massachusetts, led by Bruno Rossi and Riccardo Giacconi, that actually detected the first nonsolar x-ray source. In 1957, space research in the United States was boosted by the shock of learning that the Soviet Union had put a satellite (Sputnik) into orbit;

the West did not, at the time, think that the Soviets had such an advanced missile capability. Within a year, the United States created a civilian space agency, NASA, to promote space research, with a large degree of independence from the military's own space programs.

Wind from the Sun

It has been known for hundreds of years that a comet's tail always points away from the Sun, no matter where in its orbit the comet happens to be. The long tail seems to act like a wind sock at a weather station, billowing out to show the direction of the wind. In the early 1950s the prevailing theory to explain this effect was that "radiation pressure" from the solar light was the cause. Albert Einstein had shown at the start of the twentieth century that photons of light carry momentum, just as do high-speed particles. If there is enough light radiating out from the Sun, then it will push the comet tail downwind, causing it to point away from the Sun. This process has, in fact, been proposed as the propulsion method for deep space probes, using a "solar sail."

However, the German astrophysicist Ludwig Biermann calculated that the radiation pressure from the Sun was far too small to produce the observed effect. He noted that there is at all times a persistent weak aurora at high latitudes over the Earth, and there are also persistent small variations detected in the Earth's magnetic field. The idea was developing that these effects were due to a continual bombardment of the Earth by charged particles streaming out from the Sun. Biermann asked whether such a particle stream could be responsible for the comet-tail effect, and he calculated that a high-speed outflow, at a speed of roughly 300–600 miles per second, and a density of about 1,000–10,000 particles per cubic inch, would do the job. The outflow would represent a substantial portion of the coronal mass, which would have to be replenished from the photosphere daily, and the Sun as a whole would be constantly losing mass. However, the rate is small enough that the Sun would lose only a tiny fraction of its mass over its 10-billion-year lifetime.

This idea was not immediately accepted, because there was no obvious theoretical explanation for such an outflow. This was provided in 1957 by the work of Sydney Chapman, who showed that a hot corona will extend far out into space, and by Eugene Parker, who showed that a hot corona will expand outward continually. Chapman showed that the coronal plasma, because of the high mobility of its electrons, has the ability to conduct heat readily over long distances. If the part of the corona near the Sun is hot, he showed that the temperature will stay high far out into space, and the corona itself will extend very far away from the solar surface. Parker then calculated that such an extended corona will do more than merely reach far out into space. He predicted that the solar corona will actually expand and flow outward, much as Biermann had postulated, and he named this outflow the solar wind.

Parker's theory created a storm of controversy (he was more than once introduced at lectures with the phrase "he who sows the wind shall reap the storm"), and it was not until instruments were put into space that the high-speed particles were detected and the theory was spectacularly confirmed. Many of the earliest solar observations from space were carried out by Soviet groups, such as that of S. L. Mandel'shtam at Moscow's Lebedev Physical Institute and Konstantin Gringauz at the Radio Engineering Institute. The results ranged from measurements of the solar UV and x-radiation by Sputnik-2 to detection of the solar wind by Lunik-2 in 1959.

However, these early results were rudimentary and not widely publicized in the West. It required a more sophisticated instrument, flown by Bruno Rossi's group on Explorer-10, and then a much longer and more complete set of measurements taken on NASA's first successful planetary mission, Mariner-2, by Conway Snyder and Marcia Neugebauer in 1962, before the existence of the solar wind was fully acknowledged.

The solar wind is now being studied by a quartet of Earth-orbiting satellites known as Cluster. They were launched in 2000, replacing a similar quartet that were on a rocket that failed at launch a few years earlier. The satellites are flying close to each other in a tetrahedral (triangular-pyramid) formation so that scientists can construct three-dimensional maps of the solar wind as it passes by. The individual satellites are named Samba, Salsa, Rumba, and Tango. Cluster is studying the physical processes that take place between about 11,800 miles (19,000 kilometers) and nearly 74,000 miles (119,000 kilometers) above Earth. The project is a joint one between the European Space Agency and NASA.

The Difference It Makes

The photosphere, or visible surface of the Sun, looks fairly blank and uniform to the naked eye (through a suitable dark filter or when haze sufficiently dims the Sun, as happens most often when the Sun is low in the sky just before sunset). Those with very good eyesight, however, can occasionally see particularly large sunspots. Through a small telescope sunspots are easily visible, and are best viewed by projecting the image onto a screen. Figure 6.2 shows how much difference it makes to have the x-ray wavelengths available.

The first image in this figure shows the solar surface as it might appear from a ground-based telescope (actually, this photo was taken in the ultraviolet by the TRACE satellite, as was the second of this pair, but it shows pretty much the same solar surface that you see in white light, and it has the advantage of being simultaneous, coaligned [pointing in the same direction], and at the same magnification). The sunspots are easily visible, and in fact there is a group of sunspots in the lower right-hand portion of the image. With a good telescope, placed at an exceptionally good observing site, sunspots show a characteristic detailed structure, as seen in the first panel of Figure 6.2. The center of the spot, the umbra, is quite dark, and it is surrounded by an iris-like penumbra. The rest of the surface is covered with the fine mottling called granulation, which is caused by convective motions (like those of water boiling in a pan) of the surface. In this image you can also see something not usually found in white-light images: there are patches of brightness near the sunspots, and these are found to correlate very closely with the corona above the surface. Both the sunspots and the bright patches are caused by magnetic fields breaking through the solar

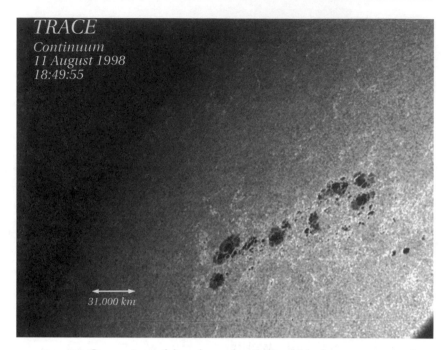

TRACE
Continuum
11 August 1998
18:49:55

31,000 km

FIGURE 6.2   Comparison of the solar surface (left), which shows sunspots and granulation, with a nearly simultaneous image of the hot corona (right) taken at a wavelength of 171 angstroms, which shows the solar atmosphere above the surface. Sunspots are places where strong magnetic fields from inside the Sun break through, and the hot corona traces out the magnetic field above the surface.

surface from inside the Sun. The second photo shows the hot corona above the solar surface, and it is fairly obvious that you would not have suspected, from looking at the photosphere, that the atmosphere above it is so rich in structure. The coronal image is noticeably different, showing large looplike structures and long thin rays radiating out of the frame. All of these features are at temperatures of several million degrees and the material filling these structures is a plasma, an ionized gas.

The x-ray light that we see coming from these regions is emitted mainly by electrons making a transition from one energy level in an atom down to a lower level. In this case, the transitions are from levels not usually seen on Earth. For instance, the right panel of Figure 6.2 shows emission from a spectral line of Fe IX, which is 8-times ionized

Fe IX 171
11 August 1998
18:48:31

31,000 km

iron. (The spectrum of neutral, or zero-times ionized, iron is labeled
Fe I, so the two counts differ by one.) One does not normally encounter temperatures of millions of degrees on Earth, except in specially
produced lab experiments and in thermonuclear explosions.

THE ARCHITECTURE OF A SPACE MISSION

If you are an amateur astronomer, it may be difficult to picture the
problems associated with putting observing equipment in space. On
the ground your basic procedure is: set up the telescope, point it at the
desired object, put your eye to the eyepiece, and observe. In space,
things are different:

1. First, there is no place to set up the telescope. It is floating
weightless in a near vacuum.

2. Next, you cannot use your hand to push the tube into position
since you are not there.

3. Likewise, you cannot put your eye to an eyepiece, since it is sev-

eral hundred miles, or farther, above your head, in a vacuum, moving at tens of thousands of miles per hour.

These are only the most obvious difficulties. Your instrument is also reaching temperatures of several hundred degrees on the side facing the Sun and is approaching absolute zero on the antisunward side. In order to get into space, it had to survive tremendous acceleration and vibrational forces from the rocket motors, which would severely damage an ordinary terrestrial instrument. If you have a computer onboard, its processor is probably overheating, because the cooling fan is useless in a vacuum, and it will fail soon anyway, because cosmic ray damage will destroy the electronic circuitry.

Sending equipment into space therefore involves difficulties that we do not normally encounter in ground-based operations. A typical space observatory must have these components:

- The scientific instruments one wishes to use.
- A "spacecraft," which is a platform that typically provides power and telemetry to the instruments, as well as the ability to point in the commanded direction.
- A launch vehicle to get these items into space.

Sounding Rockets

Most of the steps involved in putting a satellite into orbit are also found in the quicker and simpler sounding rocket launch, so we will first discuss these, before going on to satellites. Figure 6.3 shows a recent sounding rocket payload as an example of a typical launch configuration. Sounding rockets are sent up above the atmosphere for brief periods of time—typically a few minutes—but have the same set of difficulties to overcome, since they must supply their own power, pointing, and radio contact with the ground (telemetry), plus they should also have a recovery system if one wants to refly the instrumentation.

In a typical sounding rocket flight, we go through the same steps needed to put a usable satellite telescope into orbit, except that everything happens in a few minutes, instead of days or weeks. The reason,

22"

PAYLOAD

17"

BLACK BRANT VC

18"   TERRIER Mk-/0

FIGURE 6.3   Drawing of a recent sounding rocket payload used for solar studies. The two stages of rocket motors are at the bottom; the telescope (payload) is in the middle, and the power, pointing, telemetry, and recovery sections are at the top.

of course, is that a sounding rocket flight lasts only a few minutes, so there is no time for a leisurely checkout and turn-on. The design philosophy changes accordingly. For example, the skin of the rocket becomes tremendously hot during ascent owing to air friction, and then the side of the instrument facing the Sun becomes very hot in the vacuum because there is no air to take the heat away. In a satellite, the instrument is designed to reach an equilibrium condition, a balance between heating and cooling that may take hours to achieve. In a sounding rocket, the instrument is instead designed so that the heat

Table 6.1 Flight day checkout procedure and event timeline.

Time	Event	Mach number
T −240 min	Begin experiment evacuation	
T −90	All personnel at stations	
T −80	Tower function checkout	
T −65	Close vacuum valve	
T −25	Go for launch	
T −15	Verify data recorders functional	
T −5	Experiment and telemetry on internal power	
T −4	Verify telemetry contact	
T −2	Programmed launch hold	
T −0	Launch	0
T +0.6 sec	Rail release	0.1
T +6.2	First stage burnout	1.7
T +12	Second stage ignition	1.4
T +18	S-19 decouple	2.1
T +30	Verify experiment vacuum valve open	
T +44.4	Second stage burnout	6.4
T +50	Experiment cameras on	
T +61	Payload despin	6.7
T +64	Nose Tip eject	6.8
T +66	S-19 gyro caged	6.8
T +68	Attitude Control System on	6.8
T +89	Verify experiment detectors on	
T +95	Sun acquired	
T +96	Verify pointing	
T +98	Start of data-taking	
T +268	Apogee	0.4
T +457	Loss of pointing; end data-taking	5.6
T +506	Mach 1	1
T +510	Parachute open	0.03

takes as long as possible to get into the instrument; the observations are over before the telescope knows what hit it.

As an example of the steps involved in a countdown, launch, and recovery, Table 6.1 gives a partial listing of events involved in the launch and recovery of our Tuneable X-ray Imager telescope.

The prelaunch activities include a checkout of all systems and a "vertical" test, in which it is verified that signals can be sent to and from the rocket. At T −5 minutes, the umbilical lines to the payload are disconnected and the rocket goes onto internal battery power for the remainder of the flight. It is interesting to note that the famous countdown to zero that you see any time a rocket is launched is not used merely for dramatic effect. As a participant in the launch, you actually need this method in order to focus your concentration on the crucial set of steps that must be carried out in strict accordance with the timeline checklist. It is quite useful to have someone counting for you, so that you can carry out your part of the checklist without having to worry about looking at the clock. In addition, to paraphrase Samuel Johnson, the approach to zero serves to wonderfully concentrate the attention.

Note that in the case of this sounding rocket, the first stage burns for only about 6 seconds and that the vehicle accelerates to nearly twice the speed of sound (twice the Mach number) in that time. Clearly there are fearsome g-forces involved (1 g is the acceleration due to gravity that we feel on Earth as our weight; a spacecraft accelerates at several gs), but an even more difficult problem is vibration. Equipment will be literally shaken to pieces during these few seconds if it is not properly designed. At the very least, precision optics can be thrown out of alignment, electronic circuit boards can crack, or a loose screw left in the payload can come crashing through a vital piece of hardware. (In one case, an experimenter discovered after recovery that a bug had crawled into the payload; after the flight, it was found flattened against the rear bulkhead, as if someone had stepped on it.)

During first-stage burn the vehicle is spun up for stability. When the first stage burns out, it separates from the rest of the rocket and drops away. The second stage then ignites and brings the payload up to a speed of about Mach 7 (7 times the speed of sound where the rocket is), then it too burns out and separates from the payload. Once up above the atmosphere, about a minute after launch, the payload is in free fall, following a ballistic trajectory upward to some maximum height, then falling back down into the atmosphere. The payload is then despun so that it can point at the Sun and make unsmeared observations; gas jets swing it around to point at the Sun; and the attitude control system (ACS) takes over for precise solar pointing. Fortunately, the Sun is the brightest object in the sky and it is quite circular. A fairly simple sensor can therefore find the Sun and point to the center of the disk with good reliability and high precision.

All of these stages happen in under 2 minutes! The total amount of time "at altitude" for data-taking is about 6 minutes ($457 - 98 = 359$ seconds). By coincidence, this is about the duration of the longest total eclipse, so the experience for both is similar: it is all over almost before you have had time to think. In fact, one of the most important lessons that a rocket experimenter or an eclipse observer must learn is that there will not be any time to think during the observation period. Every move must be planned ahead of time, and responses to all of the possible emergencies that could arise—for instance, finding that the pointing system has not pointed correctly at the Sun—must be planned and practiced in advance.

The *most* important lesson that space experimenters must learn is that once the instrument is closed up for launch they will never be able to touch it again. This means that they must make absolutely certain that it will work as expected without further adjustments or tinkering (except for a few things that they can do by remote telemetry command). In the case of a sounding rocket, the instrument needs to work for only 5 minutes, but experimenters had better be sure that it works for the right 5 minutes. For a satellite, it must work for months or years, under very harsh conditions. While these considerations may seem obvious, a mindset very different from that of the laboratory worker is required in order to put these warnings into practice. This explains why there is a special job title called aerospace engineer.

A technique for imaging the corona in soft x-ray light, known as multilayer coating, was originally tried out on sounding rockets, and has now been used in satellites. In this method, a series of extremely thin and very precise layers are deposited onto a mirror to make it reflective to x-rays. The advantage of a sounding rocket is that it is a relatively fast and inexpensive way to get instruments into space, and to demonstrate that a new technique will work, as was done for this new imaging method. The disadvantage of rockets is that you get only

FIGURE 6.4   The TRACE satellite, shown mounted in the Pegasus launch vehicle just before the Pegasus carried TRACE into orbit in 1998. NASA systems engineer Joe Burt smiles approvingly.

about 5 minutes of observing time. However, this is infinitely better than no observing time.

Satellites

Satellites are an order of magnitude more expensive than sounding rockets, but they can carry out observations for many months or even years. One can therefore plan a far more extensive program of scientific studies with a satellite, including coordinated observations with other satellites and ground-based observing facilities. Figure 6.4 shows the TRACE satellite, which uses the multilayer technique for imaging various wavelengths of coronal radiation. It was launched on 2 April 1998 and is expected to remain functional and in its proper orbit for at least 5 years from that date. The orbit in this case is "polar sun-synchronous," which means that the satellite moves in a north-south direction, from one pole of the Earth to the other, and that the plane of the orbit faces the Sun at all times.

This type of orbit has the advantage that it is possible to have continuous, 24-hour-per-day, viewing of the Sun. A disadvantage is that the satellite passes through the Earth's radiation belts (see Chapter 8), and that produces noise in the data and can harm the electronics onboard; but the advantage of uninterrupted viewing of dynamic changes and flare events in the corona greatly outweighs the negatives.

There is also a subtlety to the orbit: if the satellite is injected into a straight up-and-down (90-degree) orbit, the plane of this orbit will remain fixed in space. As the Earth moves around the Sun during the year, the satellite will no longer point at the Sun. It will instead remain pointing in a constant direction relative to the fixed stars, while the Sun will appear to move completely around the Earth, again relative to the fixed stars. The satellite is therefore launched into an orbit that is slightly more inclined to the poles (about 98 degrees) and the bulge of the Earth's equator causes a very slow drift in the plane of the orbit (known as a precession) so that the orbit stays locked onto the sunward direction.

FUTURE MISSION PLANS: THE SUN-EARTH CONNECTION

NASA's Office of Space Science (http://www.hq.nasa.gov/office/oss/) has developed four principal science themes, including the Sun-Earth Connection (SEC) program. The SEC includes the scientific disciplines of solar and heliospheric physics, magnetospheric physics, and aeronomy, which is the study of the upper atmospheres of the Earth and other planets. The main focus of this program is to study the physical processes that link the Earth to the Sun, but it now also includes the study of other planets in the solar system as well as the study of the outer reaches of the Sun's influence into interstellar space.

The SEC program, which is connected with a major NASA initiative named Living With a Star—namely, our star, the Sun—is divided into four "Quests" for mission planning purposes:

Quest 1. How and why does the Sun vary?
Quest 2. How do the Earth and planets respond to solar variability?
Quest 3. How do the Sun and Galaxy interact?
Quest 4. How does the Sun affect life and society?

The process of developing a plan for actual space missions includes input from the scientific community, the formation of science working groups to develop detailed descriptions of proposed missions, additional higher-level scientific panels to narrow down the list to a manageable number, discussions within NASA and with Congress and the Office of Management and Budget (OMB) to set priorities, leading eventually to an Announcement of Opportunity issued by NASA. The last is an invitation to scientists to propose scientific investigations for a particular mission, and these proposals are then peer-reviewed by panels of scientific and technical reviewers to determine the winning proposals.

All of this clearly takes a long time, and it can be many years from the time a scientist thinks of a study that ought to be carried out until an instrument is flown to acquire the needed data. NASA has recently completed a review of the SEC plans, called the SEC Roadmap: Strategic Planning for the Years 2005–2020. The plan may be viewed at http://umbra.nascom.nasa.gov/spd/secr/. The Living With a Star initiative has a Website at http://sec.gsfc.nasa.gov/lws.htm. The Sun-Earth Connections Program involves activities with impact far beyond NASA's own programs. The range of implications from this program may be seen in Figure 6.5.

At any given moment, there are active space missions whose satellites are operational and sending data back to Earth, as well as missions under construction and being readied for launch. In addition, there are missions in the planning stages, either approved and in the design phase, or under consideration and not yet started. In this section we de-

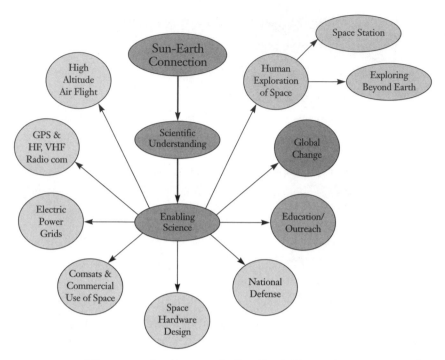

FIGURE 6.5   The connection between the Sun-Earth Connection program and areas as diverse as electric power distribution to manned space flight are shown in this chart. GPS is the Global Positioning System, a network of satellites used not only for military, commercial, and research purposes but also for general uses such as navigation systems for hikers and cars. HF (high frequency) and VHF (very high frequency) are bands of the radio spectrum often used for radio communication.

scribe the last categories—missions nearing completion, and mission ideas for the more distant future.

HESSI

The most recent solar satellite to be launched in the Small Explorer series is designed to study the particle acceleration and energy release in solar flares. The High-Energy Solar Spectral Imager (HESSI, shown in Figure 6.6) is doing this by concentrating on the energetic electrons and protons produced in flares, through observations of the x-rays and gamma rays that these particles produce. The range of energies de-

Front Grids

Solar Panels
(1 of 4)

Spin Rate
12-20 rpm

Germanium
Spectrometer

FIGURE 6.6   The HESSI spacecraft.

tected by HESSI extends from x rays as soft as 3,000 electron volts (eV) up to gamma rays at 20 million eV.

The HESSI mission consists of a single spin-stabilized spacecraft— pointing at the Sun is maintained by a gyroscope-like effect, from spinning of the entire instrument around the axis pointing to the Sun—in a low-altitude orbit inclined 38 degrees to the Earth's equator. The only instrument onboard is an imaging spectrometer—one that provides both an image and a spectrum at the same time—with the ability to obtain color movies of solar flares in x-rays and gamma rays. The word "color" here is used in the broad sense of being able to separate different wavelengths in the image.

HESSI has the finest spatial and spectral resolutions of any hard x-ray or gamma ray instrument ever flown in space. The mission is described in more detail on the HESSI Web page (http://hessi.ssl.

berkeley.edu/). The launch date of HESSI was planned to be in 2000, but damage incurred during testing delayed the launch until early 2002. It made it up in time for the end of the peak of the current solar activity cycle. HESSI is benefiting from, and providing support for, existing solar missions such as SOHO and TRACE. The planned operational life of HESSI is 3 years, although if it continues beyond that time it might operate during the Solar-B mission, with mutual benefit to both.

IMAGE

Imager for Magnetopause-to-Aurora Global Exploration (IMAGE) is a MIDEX (middle-sized spacecraft in the Explorer series) mission, selected by NASA in 1996, to study the global response of the Earth's magnetosphere to changes in the solar wind (Figure 6.7). The IMAGE scientific objectives are:

Identify the dominant mechanisms for injecting plasma into the
    magnetosphere on substorm and magnetic storm time-scales.
Determine the directly driven response of the magnetosphere to so-
    lar wind changes.
Discover how and where magnetospheric plasmas are energized,
    transported, and subsequently lost during substorms and mag-
    netic storms.

In order to fulfill its science goals, IMAGE utilizes neutral atom, ultraviolet, and radio imaging techniques. A suite of three neutral atom imagers (NAI) provides energy- and composition-resolved images at energies from 10 eV to 200 keV with a time resolution of 300 seconds. Two ultraviolet imagers, covering wavelength ranges from 120–180 nm (FUV) and 30.4 nm (EUV), provide coverage in the far and extreme ultraviolet. A radio plasma imager (RPI) transmits and receives pulses from 3 kHz to 3 MHz, allowing relative motions of the satellite and plasma to be determined to a resolution of 400 m/s and a time resolution as good as 4 s. The IMAGE satellite was launched on 25 March

FIGURE 6.7   The IMAGE spacecraft during prelaunch ground testing.

2000. More information on this mission is available at: http://image. gsfc.nasa.gov/.

TIMED

The Thermosphere • Ionosphere • Mesosphere • Energetics and Dynamics (TIMED) Program is the first science mission in the Sun-Earth Connection Program as detailed in NASA's Strategic Plan. TIMED (Figure 6.8) will explore the Earth's mesosphere and lower thermosphere (60–180 kilometers high), the least explored and understood

FIGURE 6.8   Artist's concept of the TIMED spacecraft.

region of our atmosphere. It is known that the global structure of this region can be perturbed during stratospheric warmings and solar-terrestrial events, but the overall structure and dynamics responses of these effects are not understood. Advances in remote sensing technology employed by TIMED instrumentation will enable us to explore this region on a global basis from space.

The primary objective of the TIMED mission is to investigate and understand the energetics of the mesosphere and lower thermosphere/ionosphere (MLTI) region of the Earth's atmosphere. The MLTI is a transition region in which many important processes are known to change dramatically, yet there are only sketchy and incomplete measurements of the region. This is where temperature variations, which increase dramatically with altitude, are largest, and where atmospheric motions are dominated by poorly understood gravity waves and tides. When the mission is complete, it will have:

Performed the first focused exploration of the MLTI region—a major link in the solar-terrestrial chain.

Traced the flow of energy and momentum from the lower into the upper atmosphere.

Quantified the dramatic influences of the Sun and outer space on the Earth's middle and upper atmosphere, increasing our ability to predict space weather.

Examined the major energy sources, transport processes, and energy sinks in the MLTI, and applied this information to define the channels by which energy enters into, travels through, and is lost from the region.

Supplied knowledge of the upper atmospheric circulation pattern and its role in the transport of chemicals important in ozone chemistry.

As of this writing, the TIMED launch date is early 2001. Additional information about the mission can be found at TIMED's Website, http://www.timed.jhuapl.edu/home.htm, from which the above description was obtained.

STEREO

One of the missions in the line of solar-terrestrial probes is STEREO (Plate XII). The goal of the STEREO mission is described on NASA's announcement page as follows:

The Solar TErrestrial RElations Observatory (STEREO) mission is the third in the line of Solar-Terrestrial Probes and is a strategic element of the Sun-Earth Connection Roadmap. The Solar-Terrestrial Probes program is a continuous sequence of flexible, cost-capped missions designed for sustained study of critical aspects of the Sun-Earth system. It is an outgrowth of the highly successful International Solar-Terrestrial Physics program.

The primary goal of the STEREO mission is to advance the understanding of the three-dimensional structure of the Sun's corona, especially regarding the origin of coronal mass ejections (CMEs), their evolution in the interplanetary medium, and the dynamic coupling between coronal mass ejections and the Earth environment. Coronal mass ejections are the most energetic eruptions on the Sun, are the primary cause of major geomagnetic storms, and are believed to be responsible for the largest solar energetic particle events. They may also be a critical element in the operation of the solar dynamo because they appear to remove dynamo-generated magnetic flux from the Sun.

STEREO will, therefore, continue the systematic study of the relationship between processes on the Sun and their consequences for the Earth. The ejections of well-defined clouds of plasma from the corona were discovered in 1973 by instruments on NASA's Skylab/Apollo Telescope Mount. Although studies continued with the NASA Solar Maximum Mission and the ESA-NASA Solar and Heliospheric Observatory (SOHO), these investigations were limited to vantage points that best showed those coronal mass ejections that missed the Earth. However, with two spacecraft sent in opposite directions away from the Sun-Earth line, the STEREO mission will finally allow unambiguous observations of those coronal mass ejections that directly impact the Earth. STEREO will also for the first time provide a stereoscopic view of the three-dimensional corona and the interplanetary medium and thereby advance the Sun-Earth Connection understanding of the heliosphere. See http://sd-www.jhuapl.edu/STEREO/.

Solar-B

Solar-B is a mission of the Japanese space agency proposed as a follow-on to the highly successful Japan/U.S./U.K. Yohkoh (Solar-A) collaboration, and has major American participation that includes one of us. The mission (Plate XIII) consists of a coordinated set of optical, EUV, and x-ray instruments that will apply a systems approach to the interaction between the Sun's magnetic field and its high-temperature, ionized atmosphere. The result will be an improved understanding of the mechanisms that give rise to solar magnetic variability and how this variability modulates the total solar output, creating the driving force behind space weather.

Solar-B will, for the first time, provide quantitative measurements of the full vector magnetic field on small enough scales to resolve elemental flux tubes. The field of view and sensitivity allow changes in the magnetic energy to be related to both steady state (coronal heating) and transient changes (flares, coronal mass ejections) in the solar atmosphere. Solar-B is an excellent opportunity for highly leveraged U.S. participation in a major mission that will greatly advance our understanding of the crucial first link in the Sun-Earth Connection. Solar-B started in 1999, will be launched in 2005, and should operate for at least 3 years, with a design life of 7 years.

The science objectives of the Solar-B mission are:

1. Understand the creation and destruction of the Sun's magnetic field: Magnetic fields permeate all space and play an important role in shaping the universe on all scales. The fields are continuously being generated by dynamos in stellar interiors and swept out into space by stellar winds. The solar dynamo is sufficiently near and operates on a short enough period, 11 years, that it can be studied directly.

2. Study the modulation of the Sun's luminosity: During the last decade observations from space have led to the profound discovery that the total output of energy from the Sun is not constant but varies in phase with the magnetic activity cycle. The amplitude of this variation, over the single cycle measured, though small was only a factor of three to five below the level required for a significant climatic response. Solar-B will make the first observations with resolution, wavelength coverage, and time span adequate to determine the mechanism for the magnetic modulation of solar luminosity.

3. Understand the generation of UV and X radiation: The Sun is a powerful and highly variable source of ultraviolet, x-rays, and energetic particles, which are known to have major effects on our environment. This high energy radiation must be due to the annihilation of magnetic energy in the Sun's atmosphere, the chromosphere and corona. Due to its broad complement of instruments with high spatial and spectral resolution, Solar-B will be able to study processes such as magnetic reconnection and wave dissipation that are believed to be responsible for the conversion of magnetic energy into UV and X radiation.

4. Understand the eruption and expansion of the Sun's atmosphere: The million-degree corona continually expands outward, becoming a supersonic wind that blows past the Earth, buffeting the geomagnetic field and energizing the upper atmosphere. In addition, large parts of the corona are seen to erupt, blasting through the solar wind and causing major magnetic disturbances at Earth. Solar-B will provide accurate measurements of magnetic fields, electric currents, and velocity fields that supply the corona, thus revealing the root causes of the Sun's eruptions.

This description, and additional information, can be found at the Solar-B home page at the Marshall Space Flight Center: http://www.ssl.msfc.nasa.gov/ssl/pad/solar/.

In order to meet the scientific objectives stated above, the satellite will carry the following instruments:

1. The SOT: a 0.5 meter (20-inch) solar optical telescope equipped with a vector magnetograph, narrow-band imager, and spectrometer to obtain photospheric magnetic and velocity fields at 0.2 arc seconds (150 kilometers) resolution.

2. The XRT: an x-ray telescope to image the coronal plasma in the range 1–20 million degrees at 2 arc seconds resolution and a large field of view.

3. The EIS: an EUV imaging spectrometer to obtain plasma velocities to an accuracy of 10 km per second along with temperatures and densities in the transition region and corona at 2 arc sec resolution.

The X-ray Telescope (XRT)

The basic goals of a soft x-ray telescope for the Solar-B mission are to facilitate the study of the dynamics of fine-scale coronal phenomena, such as magnetic reconnection and coronal heating mechanisms, while at the same time recording the large scale global phenomena, such as coronal mass ejections. In order to meet these objectives, the x-ray telescope (XRT) (Figure 6.9), the instrument for which one of us (LG) is the principal U.S. investigator, will work closely with the focal plane

Figure 6.9   Drawing of the x-ray telescope, shown mounted under the optical telescope being built for the Solar-B mission.

instruments of the solar optical telescope (SOT) and with the EUV imaging spectrometer (EIS).

The U.S. and Japanese Solar-B science teams have identified key problem areas to be addressed by this mission. Among those in which the x-ray telescope plays a major role are:

1. Flares and coronal mass ejections. How are they triggered, and what is their relation to the numerous small eruptions of active region loops? What is the relationship between large-scale instabilities and the dynamics of the small-scale magnetic field?

2. Coronal heating mechanisms. How do coronal loops brighten? Are wave motions visible, and are they correlated with heating? Do loops heat from their footpoints upward, or from a thin heating thread outward? Do loop-loop interactions contribute to the heating?

3. Reconnection and coronal dynamics. Yohkoh observations of giant arches, jets, kinked and twisted flux tubes, and microflares imply that magnetic reconnection is significant in coronal dynamics. With higher spatial resolution and with improved temperature response, the XRT will help clarify the role of reconnection in the corona.

4. Solar flare energetics. Although Solar-B will fly during the next solar minimum, there will still be many flare events seen. The XRT is designed so that it can test the hypothesis about magnetic reconnection as a source of energy release that has emerged from the Yohkoh data analysis.

5. Photosphere/corona coupling. Can a direct connection be established between events in the photosphere and a coronal response? To what extent is coronal fine structure determined at the photosphere?

In order to address these scientific problems, the x-ray telescope (XRT) on Solar-B will have a grazing-incidence mirror. It will be similar to that flown on Yohkoh, but with several improvements:

- The pixel size, which determines how fine the detectable detail is, improves from 2.46 arc seconds to 1.0 arc second.
- The detector resolution is increased from 1,024 × 1,024 pixels to 2,048 × 2,048 pixels with full-Sun field of view;
- The temperature sensitivity is augmented with a low-temperature (1–2 million kelvins) response in addition to a high-temperature response.

In comparison with the Yohkoh soft x-ray telescope (SXT), the Solar-B x-ray telescope (XRT) will therefore have six pixels for every 1, a comparable field of view, comparable temperature diagnostic accuracy, and a greater range of temperature coverage.

The EUV Imaging Spectrograph (EIS)

The EIS consists of a multilayer-coated single mirror telescope and a stigmatic imaging spectrometer incorporating a multilayer-coated diffraction grating. What this last term means is not only that the instrument will form a spectrum, but also that there will be a long narrow slit, along which an image of the Sun at EUV wavelengths is formed, while a spectrum from each point along the slit is formed in a direction perpendicular to the slit. One therefore obtains EUV spectra of a slice of the Sun, rather than of only a single small box.

The image is produced by the primary mirror, which focuses the light onto a slit. On the other side of the slit, there is a curved diffraction grating, which is positioned such that it takes the light coming through the slit and diffracts it back onto a detector. The light that passes through this spectrometer aperture—which can either be a narrow slit or a thin rectangular box, to provide actual images—is dispersed and reimaged as a focused image onto the CCD detector.

The Solar Optical Telescope (SOT)

The largest instrument in Solar-B is the Solar Optical Telescope (SOT), which has a package of instruments, the Focal Plane Package (FPP), located at the optical telescope's focus (Figure 6.10). The optical telescope is a mid-size optic, 0.5-meter (20-inch) diameter primary, of Cassegrain design. This arrangement gives a long focal length, so that the image of the Sun at the focal plane is large, and the CCD camera pixel size corresponds to 0.1 arc seconds of angular resolution. Although it is sometimes possible to obtain better resolution from the ground, the satellite will provide continuous, uninterrupted, and uniform-quality data, which cannot be obtained from the ground.

The FPP is a sophisticated package that provides high-resolution images of the photosphere and chromosphere at a number of different wavelengths. Most significant, it provides magnetic field measurements of unprecedented quality. In particular, extreme care is being taken to build an instrument that can provide not only the strength of the magnetic field, but its direction as well. This is known as measuring the vector magnetic field, and the data derived may be valuable in making the comparison between the field that penetrates the solar surface and the observed structure and activity of the corona above.

Solar Probe

The Solar Probe is designed to provide direct sampling of the solar atmosphere by actually flying through it (Figure 6.11). It is a unique high-technology dual spacecraft mission to be implemented in cooperation with the Russians. It will venture deep into the solar corona, the

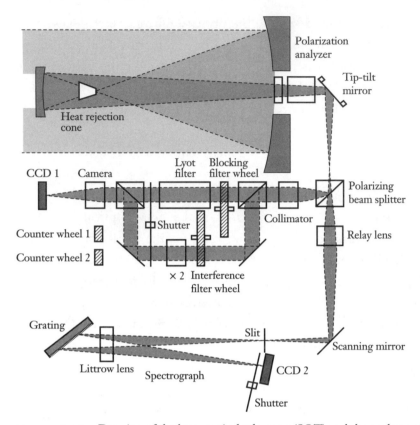

FIGURE 6.10 Drawing of the large optical telescope (SOT) and the package of focal plane instruments (FPIP) being built for the Solar-B mission.

Sun's outer atmosphere—far closer to the Sun than any other spacecraft has previously ventured. It will demonstrate technologies for imaging, and take *in situ* and remote measurements at 3 solar radii above the solar surface, where radiation temperatures exceed 2,000 K.

Solar Probe, technically part of the Outer Planets program at NASA, is designed to answer long-standing fundamental questions:

- What heats the extended solar corona and accelerates the solar wind?
- What are the sources of the solar wind in the corona?
- What roles do turbulence and waves play in the coronal heating process?

FIGURE 6.11   Artist's conception of the Solar Probe nearing the Sun.

• What are the mechanisms that accelerate, store, and transport high-energy particles throughout the corona?

Our Sun is the only star we can measure *in situ*. Solar Probe will relate remotely observed solar phenomena to actual physical processes occurring in the solar atmosphere and corona. These findings will offer unique access to processes that are important in many other significant but inaccessible stars throughout the Universe.

A conceptual view of the Probe satellite near its minimum approach to the Sun is shown in Figure 6.11. The description provided above, as well as additional information about the proposed mission, can be found at: http://www.jpl.nasa.gov/ice_fire/sprobe.htm. As currently planned, the Solar Probe will be launched toward Jupiter in February 2007, arrive at Jupiter in June 2008, and arrive at perihelion, the closest point to the Sun in its orbit, in October 2010.

Between Fire and Ice

From the point of view of climate, the Earth is basically a large rock floating in cold, empty space with an enormously bright searchlight shining on one side of it. If the Earth were not rotating and did not have an atmosphere, then the side being illuminated would be hotter than boiling water and the dark side would be solidly frozen. A rotating Earth would be more evenly heated, like a rotisserie chicken being cooked, but without an atmosphere the *average* temperature around the globe would still be below the freezing point of water. The warming due to the atmosphere, known as the greenhouse effect, is necessary to keep the Earth habitable.

What we call "climate" is a complex interaction between the heating of the Sun and the processes that distribute this heat over the planet. In this chapter we provide a perspective on the forces determining the overall climate of the Earth. The goal is to provide the big picture, the global climate and its relation to the solar input as an organic whole. We will then be in a position to discuss the major causes of climate variability and the role played by the Sun in climate change. As we will see, in the twentieth century the natural changes due to the Sun were overwhelmed by the human influences on global climate.

THE GENESIS OF CLIMATE

By the climate of a region, we usually mean the average, over some long time-scale, of temperature and precipitation. Since we normally omit day-to-day variability, or weather, in conceptualizing climate, the climate of a region might be thought of as the "averaged weather."[1] We use words such as warm or cold, moist or dry, but the inadequacy of these descriptions becomes apparent as soon as we ask a few questions: Do we mean summer or winter? Typical daytime or typical night-time? Do we average over a month, a year, or a century? Do we mean present-day conditions or those prevailing during the last ice age? Is the occasional killer tornado part of the climate? What exactly do we mean by climate anyway?

The local climate of a region is determined by many factors: Is there a body of water nearby? Is the region downwind of a mountain chain? What is the pattern of ocean currents on the usual storm tracks? What is the typical composition of the soil? Is there a large urban area nearby, producing excess heat and air pollutants?

What we really need to speak about is the climate system. Climate is determined not only by what the air is doing, but also by the interaction of air, sea, land, and ice, forming one interconnected whole. Changes in one part affect the others and are in turn affected by them. What we have is a complex, dynamic linking of the atmosphere, the oceans, lakes and rivers (hydrosphere), the frozen water masses (cryosphere), the underlying ground (lithosphere), the plant and animal life

1. Hubert Lamb, founder of the Climatic Research Institute at the University of East Anglia, and one of the inspirational forerunners in the field, considers the term "averaged weather" to be incorrect since we must specify "not only averages but the extremes and the frequencies of every occurrence that may be of interest." In his book *Climate, History, and the Modern World* (1995) he suggests that climate be defined as "the total experience of the weather at any place over some specified period of time." His objection is a real one—the range of variability in the short term at most locations is so large that averaging can be quite misleading. Moreover, the changes in "average" climate from one millennium to another, or even from one century to the next, are so large that the word "average" hardly seems appropriate. Yet one has the intuitive feeling that the word "climate" has meaning, and we will for the time being speak as if "typical climate" is a meaningful concept.

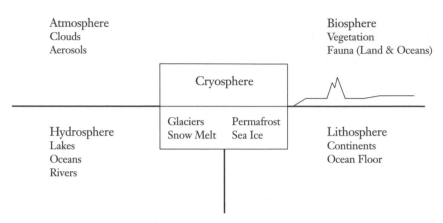

Figure 7.1 Schematic illustration of the climate system and its subsystems.

(biosphere), and the energy source that drives the interactions among all of these, the solar radiation. These systems are all interlinked by exchanges of energy, mass, and other dynamic variables to form a local climatic system, as illustrated schematically in Figure 7.1.

The basic mechanism behind all of these dynamics is differences in heating between one place on the Earth and another, followed by a redistribution of energy from warm regions to cool ones. The source of nearly all the energy driving the climate system is the Sun, and one of the goals of climatologists is to devise a model that takes this energy input, feeds it into a model Earth, and comes out at the end with the observed climate. Because the real system is quite complicated, this goal has not yet been achieved, but we can approach the subject in stages, starting with the most fundamental parts and then adding to the model bit by bit. In mathematical language, this is equivalent to starting with the largest terms in the equations and then adding additional terms as we become more familiar with the solutions.

A Simple Climate System Model

First, imagine some surface exposed to the Sun—the hood of your car on a sunny day might be a good example (see Figure 7.2). Energy from the Sun, mostly in the form of visible and infrared light, is constantly

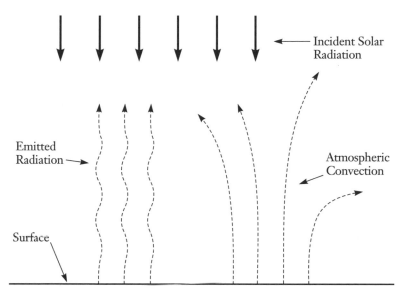

FIGURE 7.2   Heating of a surface by the balance of incoming and outgoing energy.

falling on this surface. Some of this energy is absorbed into the surface, causing it to heat. We now ask, how hot will it get?

If the surface just keeps absorbing energy it will keep getting hotter forever. Since we know that this doesn't happen, there must be some process or processes that remove energy from the surface. In this case, the two major ones are: First, air in contact with the surface is heated and then moves away, taking some of the heat with it. And second, the surface radiates energy into the surrounding space. (There are also other processes operating, such as conduction of heat into and along the metal, but this model is adequate for the moment.) The first process is familiar and needs little explanation: anything in contact with the heated surface will tend to come to the same temperature as that surface. In particular, the air around it will be heated, and movements of the air will transport some of that heat away. Note that this implies that in a vacuum the same surface, exposed to the same illumination from the Sun, will be hotter than in air. This is indeed true for objects in space or on the Moon.

The loss of heat by radiation is somewhat less familiar to most people. It is a fact that all objects radiate electromagnetic energy (light). More or less of it is radiated, depending on the temperature of the object, and the average wavelength of the radiation also depends on the temperature. At normal "room temperature" most of this radiation is in the infrared, and is therefore not visible to our eyes, but it can be detected with special infrared-sensitive cameras. (Such cameras can also be used to image homes to detect heat leaks, which can then be sealed to reduce heating costs; this is especially important in cold climates.) The reason that all objects emit such radiation is that matter is made of atoms, which consist of charged particles—a nucleus surrounded by electrons—and that charged particles emit radiation when they are accelerated.

What does acceleration of charged particles have to do with temperature? What we really mean by temperature is the motion of the atoms, or other constituent particles, in a quantity of matter. High temperature simply means that the atoms are moving rapidly. The understanding of this relation between such gross, averaged quantities as "temperature" and the detailed states of the numerous, submicroscopic particles of which matter is composed, was one of the greatest triumphs of late-nineteenth-century physics. (These studies also led to quantum mechanics in the twentieth century.) Temperature corresponds to rapidly moving atoms, which also means atoms colliding with each other, since numerous atoms moving around at high speed will unavoidably encounter one another from time to time. In some cases, when atoms are held in place so they cannot move large distances, the energy takes the form of vibrations or other periodic motions such as rotation around the fixed spot.

All of these represent types of accelerated motion, since only motion at a constant speed in a straight line is unaccelerated. Acceleration means change of motion, so that deflection via collision and oscillation in place both represent examples of accelerated motion. Typically it is the very light-weight electrons that are most easily moved and that therefore radiate most efficiently.

The amount of radiation emitted by matter increases very quickly with increasing temperature. Qualitatively, the rapid increase is understandable because, as the temperature goes up, the atoms are moving more quickly so they collide more often. The collisions are also more violent—the accelerations are greater, and larger acceleration means that the radiation is stronger. It also means that the typical photon of light emitted in the collision is more energetic—has a shorter wavelength.

Higher temperature thus leads to much stronger, shorter wavelength radiation. While objects at what we call room temperature emit their photons mainly in the infrared, for the much hotter Sun the radiation peaks in the visible (or more likely, it is "the visible" because this is where the solar radiation peaks). Hotter stars are bluer in color because their peak photon energies are even higher.

Thermal Balance

We are now in a position to understand why the surface in our example reaches a certain temperature and then does not continue to get hotter. As it absorbs energy from the Sun, its temperature increases. But conduction to the surrounding air and radiation into the surrounding space both increase with the temperature of the surface. Eventually, the surface becomes hot enough that it is losing as much energy to these two processes as it is gaining from the incoming sunlight. It has achieved thermal balance and will remain at a constant temperature so long as nothing else changes—so long as a passing cloud does not dim the Sun or the air does not become more humid, thereby changing its heat-carrying capacity, for instance.

A simple calculation of the Earth's equilibrium temperature, based only on the balance between the Sun's energy coming in and the energy radiated by the heated terrestrial surface going out (including the effects of the albedo of the Earth—the fraction of the incoming solar radiation that is reflected back out), gives about 255 K, whereas the correct value is about 33 K (60°F) higher. (On the Kelvin scale water freezes at 273 K, which corresponds to 0°C; water boils at 373 K, or

100°C). So we must account for the fact that the equilibrium temperature is about 15°C (about 60°F) instead of the −18°C (0°F) that we calculate. Clearly there is more involved than has been included in our
model so far.

Heating a Spherical, Rotating Earth

The simple example of a car hood we gave above would lead to a situation in which the surface is heated to some temperature and then remains there. While this represents the basic warming mechanism for
the Earth as a whole, there are at least three additional factors that have
to be considered in trying to model the Earth's climate:

- The Earth is a spherical body.
- The Earth is rotating.
- The Earth has an atmosphere.

Because the Earth is spherical, the latitudes near the equator face the
Sun more directly and so receive more heating than do those near the
poles, as is shown in Figure 7.3. This means that the air in the tropics
will warm and expand, while the air in the polar regions will cool and
contract. Since air is free to move if pressed, the heated air near the
equator expands, and moves outward and upward. Air from the equatorial regions then flows away from the equator poleward in both northern and southern directions. This flow eventually cools and sinks, then
flows back down toward the equator as a surface flow.

The overall pattern is two torus-shaped rolls, one in each hemisphere. This type of circulation, a pair of giant convection cells between pole and equator, is usually called the Hadley circulation, after
George Hadley, who proposed this flow pattern in 1735. But it is only a
first approximation to the truth.

We now need to understand why these atmospheric flows go *around*
the poles, rather than straight toward them. This turns out to be due to
the fact that the Earth is rotating. The first effect of the Earth's rotation
is, of course, the large change in solar illumination between day and

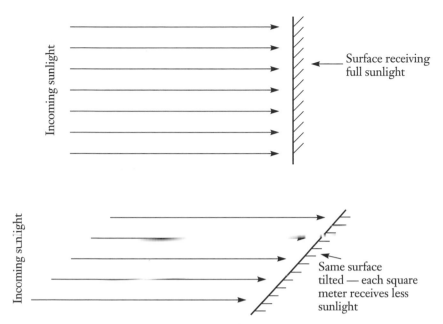

FIGURE 7.3 For a spherical Earth, different latitudes have different tilts relative to the incoming sunlight, causing differences in the effectiveness of solar heating. The tilted surface in the lower part of the figure is the same length as in the upper part, but receives less sunlight because of its tilt.

night. This causes not only an overall day/night temperature difference, but in many locations a heating/cooling of land relative to the ocean, with a corresponding shift in wind direction from day to night: the land changes temperature much more quickly than does the ocean, so that in strong sunlight the land is hotter than the water during the day, and cooler than the water at night. However, this rapid cycling of the heat input averages out and has negligible long-term effect. Much more important is the effect that the rotation has on masses of air or water that try to move.

Because the Earth is a rotating sphere, objects near the equator have a large west-to-east velocity. We can easily calculate the speed by taking the Earth's circumference and dividing by the number of hours in a day: 24,800 miles divided by 24 hours is approximately 1,000 miles per hour. Everything in contact with the surface of the Earth—houses, people, air, birds—moves at this speed in the direction of rotation. (Rockets carrying satellites into orbit are usually launched west-to-east, and as close to the equator as possible, so that they start out with this "boost.")

In contrast, an object located at one of the poles has no sideward velocity, while objects in intermediate latitudes have velocities in between zero and 1,000 miles per hour. The upshot of all this is that, when air flows north-south, it develops a relative velocity that does not match that of the surface. If equatorial air moves toward the poles, it has more west-to-east velocity than the ground over which it is passing; if polar air moves toward the equator, it has less west-to-east velocity. The air will therefore seem, to people on Earth, to curve to the side, toward the east if it is moving to higher latitudes, and toward the west if it moves to lower ones. One might say that it "overshoots" the motion of the surface if it arrives with high west-to-east velocity, and the opposite for the reverse situation.

The final wind flow pattern is determined in large part by the fact that the moving air will be pushed sideways, as if by a force. This apparent force, resulting from the rotation, is known as the Coriolis force (named after the French engineer Gustave-Gaspard Coriolis, 1792–1843, who described this effect in an 1835 paper; he also introduced the terms "work" and "kinetic energy" with their modern meanings in physics). Note that it is only as recently as 1980 that our present understanding of the large-scale circulation of the atmosphere, especially a mathematical treatment of the correct physical quantities that are conserved in the flow, was arrived at.

Figure 7.4 shows that the general circulation of the Earth's atmosphere divides into bands, with bands of high-pressure air at 30 degrees and of low pressure at 60 degrees north of the equator, as well as roughly symmetric bands at the same latitudes south of the equator. The winds flowing generally toward the west are called trade winds because sailors relied on them on their trade routes. Their location explains why Columbus had first to sail south in order to cross the Atlan-

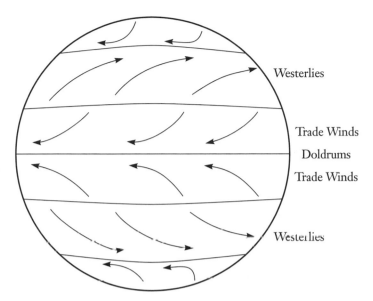

FIGURE 7.4   Typical global air circulation of the Earth.

tic and why he wound up in the Caribbean. The winds from the west
are called westerlies, for obvious reasons. Near the equator the winds
are not strongly curved by the rotational force, so that they are disorga-
nized and variable. This area, known as the doldrums, is difficult and
unreliable for sailing.

Note that there is a second band of east-to-west winds at high latitudes, above 60 degrees. These were routinely used by northern Euro-
peans to cross the North Atlantic at least as early as the ninth century
and into the fourteenth century. The global cooling between the time
of the Medieval Grand Maximum and the Little Ice Age, described
later in this chapter, caused the North Atlantic to freeze over, making crossing impossible and stranding settlers in Greenland and New-
foundland.

The Earth's atmosphere acts like a blanket to retain some of the incoming solar energy and to warm the surface. The basic mechanism of
this so-called greenhouse effect is fairly simple, although as with nearly
any scientific subject, the details can be formidable. The key point is
that our atmosphere is not equally transparent at all wavelengths. Even

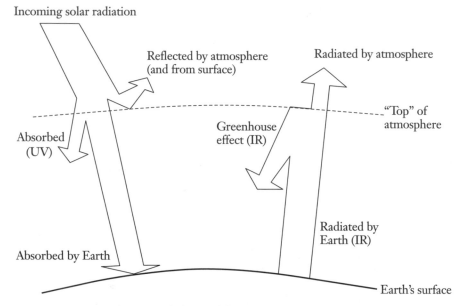

Incoming solar radiation

Reflected by atmosphere
(and from surface)

Radiated by atmosphere

"Top" of
atmosphere

Absorbed
(UV)

Greenhouse
effect (IR)

Radiated by
Earth (IR)

Absorbed by Earth

Earth's surface

FIGURE 7.5   The energy balance of the Earth.

though it seems invisible to us at the wavelengths to which our eyes are sensitive, the atmosphere is actually quite absorbing—opaque—at the longer, infrared wavelengths. Why is this an issue? Because the global temperature is determined by the balance between incoming and outgoing energy, and the incoming energy is mainly at shorter wavelengths—those we call "visible"—so that the solar energy can enter the atmosphere and heat the surface.

But the surface of the Earth and the atmosphere near the ground emit energy at much longer wavelengths, and the atmosphere does not allow that energy to radiate freely back out into space. This effect is sometimes called blanketing, and it results in a warming of the Earth's surface, which leads to an increase in the radiation back out (and a shift toward shorter wavelengths) until a new equilibrium is achieved at a higher temperature. This type of energy trap is often generically known as a greenhouse effect. It is included in Figure 7.5, which shows an overview of the way in which the incoming solar energy is absorbed, redistributed, reradiated, and reflected at various parts of the Earth's surface and atmosphere.

Our model is now improved, in that it shows day/night differences, it predicts that there will be a general circulation of the atmosphere, and it explains the basic greenhouse warming of the Earth. But we know that any given place on the Earth also has seasons, so there must be an explanation for the average difference between, for example, summer and winter. As we will see, the same factors that explain seasons also explain the far more drastic "seasons" known as ice ages.

To make the model more realistic, we now include two properties of the Earth's orbit relative to the Sun:

- The orbit is slightly elliptical, so that the distance from the Earth to the Sun varies slowly throughout the year.
- The rotation axis of the Earth, the imaginary line connecting the north pole to the south pole, is tilted relative to the plane of the Earth's orbit. Also, the angle of tilt varies by a few degrees, with a period of about 41,000 years.

The ellipticity of the orbit means that the distance between the Earth and the Sun varies throughout the year: rather than being perfectly circular, the orbit is slightly elliptical. Because of perturbations from the other planets, the ellipticity of the Earth's orbit is not truly constant, but varies from near zero to as high as 6 per cent, with a period of roughly 100,000 years. In our times the ellipticity is fairly low and it will remain small for the next 50,000 years; this may have an effect on the timing and severity of the next few ice ages, as we discuss later in this chapter. The effect of the small ellipticity does not cause the seasons, as we can verify from the fact that the Earth comes closest to the Sun on 4 January each year, in the Northern Hemisphere's winter. It is the tilt of the Earth's rotation axis that causes large differences between summer and winter, especially at high latitudes, and the larger the tilt, the larger the effect. Thus the seasons are caused by the Earth's tilted axis and not by the Earth's varying distance from the Sun. Because of the effect we showed in Figure 7.3, the Northern Hemisphere is warmed when it faces the Sun, while at the same time the Southern Hemisphere sees the Sun only at an angle. Thus Christmas, 25 December, comes in the summertime in Australia, while it is winter in the United States.

Anyone who has played with a gyroscope will remember that the wheel spins many times while the axis remains fairly steady, and this is what the Earth does as it orbits the Sun. However, a gyroscope's spin axis will also slowly change in a type of motion called precession. Likewise, the direction of the tilt with respect to the long axis of the orbit's ellipse varies, like the wobbling of a gyroscope or a top, with a period of about 21,000 years. As we will see, all of these time-scales show up in the climate record connected with the expansion and contraction of the great polar ice sheets.

THE CLIMATE RECORD

By putting together the best available evidence from all sources—tree rings, isotope ratios, sedimentation rates in core samples, and so on— we can try to get some idea of the various time-scales for climate change that the Earth has witnessed. The model we discussed in the previous section predicts that certain periodicities, or near-periodicities, will be observed, such as day-night cycles, summer-winter changes, and the longer-term changes due to variations in the Earth's orbital parameters. These are indeed observed, but there are other variability time-scales observed as well, on a time-scale of about 200 to 500 million years. The record indicates that there seem to have been long periods in the Earth's history without ice ages, interleaved with eras having dramatic climate variations and numerous, relatively rapid ice/no-ice periodicities.

The key to these long-term climate changes seems to be continental drift, now explained as part of the theory of plate tectonics. On a time-scale of 100 million years there is considerable motion of the Earth's crust, with major rearrangements in the sizes and positions of the continents. Occasionally it will happen that a large land mass finds itself at one of the poles, making it possible for a large permanent glacier to form; this is presently the situation in Antarctica. This polar ice cap has an effect throughout the entire global climate system. In particular, it permits the orbital variations to cause periodic ice ages, through a mechanism we will discuss later in this chapter, the Milankovitch Cycle. Keep in mind that what is behind all of these climate changes is a variation in the amount of solar heat retained by the Earth, and in the way this heat is redistributed among the various elements of the climate system.

First we discuss two basic issues related to the connection between the solar input and the Earth's climate: how we go about reconstructing past climates, and how we avoid concluding that the Earth should have been totally covered with ice in the past and should still be so today.

Carbon Dating

From time to time one hears news reports of the age of some ancient artifact having been established by carbon dating. What is this method of determining age? What is being measured? Why does this measurement tell us how old something is? To answer these questions we need to take a short detour into nuclear physics, explaining along the way why alchemists failed to turn lead into gold. (Alchemists, long ago, sought the Philosopher's Stone to help them with the transformation. Harry Potter's fictional encounter with the Philosopher's Stone, with its name changed to the Sorcerer's Stone by the publisher for American readers, is an exciting tale read by millions.)

Most of what we see happening around us in the world involves the shuffling around of electrons. Chemical reactions, the hardness or softness of materials, freezing and melting, the emission of light and other electromagnetic radiation, life—these are all processes in which electrons move between atoms or are in some way induced to move or to stay in place within bits of matter. Electrons are held by their attraction to the atomic nucleus, and the amount of energy needed to change or to break this bond determines whether or not such changes can happen in our typical, everyday surroundings. Clearly there is enough energy available in sunlight, or in the hydrocarbon burning that permits life to flourish on this planet, or in the batteries that power our cellular telephones.

We do not routinely see lead changing into gold, although careful scrutiny does reveal an occasional atom of uranium changing into an atom of lead. But that transformation is rare, and in general an element stays what it is, even through fire and ice. So why do elements retain their identity, even though many other changes are happening?

The atom consists of a heavy nucleus having some number of positive electric charges—the protons—surrounded by a cloud of extremely light-weight negative charges—the electrons—which are attracted to the positive charge of the nucleus. Electrons are not held very tightly to the nucleus and so there are usually some electrons moving freely through any bit of matter, and they will inevitably pass close to and be attracted by any nucleus that happens to have too few electrons around it.

What does "too few" mean? If the nucleus has attracted a number of negative charges equal to the number of positive charges, then it has no net charge and the passing electron has no excess charge to equalize. In general, the number of electrons that a nucleus attracts will be equal to the number of positive charges, so that the atom overall is electrically neutral.

The properties of elements as we see them are determined by the number of electrons in the atom. But this number is actually determined by the number of positively charged protons in the nucleus, since the nucleus will attract whatever number of electrons are necessary to make the atom as a whole have no net charge. The reason that elements retain stable properties through all of the flux of daily life is that the nuclei remain stable.

The reason nuclei remain stable is that a force much stronger than the electromagnetic force holds them together. When neutrons and protons get close enough together, they lock on to each other via the strong nuclear force. Like charges repel, so the protons resist being pushed together, and large amounts of energy are needed to push the protons close enough together to engage the strong nuclear force. (Think of a jack-in-the-box being pressed into its container, which then snaps closed to hold it in place.) But once the nuclei are hooked up, a large amount of energy is needed again to release them.

Nuclei tend to be stable at the energy levels normally encountered in our terrestrial environment. Alchemists tried to use hot fires to induce what we now know be nuclear changes, but they would have needed a fire at many tens of millions of degrees to accomplish their goal. So they were able to achieve some superficial chemical changes, such as a rearrangement of which elements were sharing electrons with other elements in a chunk of matter. But they did not come close to reaching the level needed to alter the composition of the nuclei, and the elements themselves remained unchanged.

Some nuclei are more stable than others, and those that are not stable decay—change—into nuclei of other elements. This decay can happen in several ways, almost all of which are accompanied by the ejection of a packet of energy (a photon) or some type of particle (electron, neutron, alpha, and so on) from the nucleus. Before the particles being ejected from the nuclei were identified, they were called "alpha ray," "beta ray," "gamma ray," for what turned out to be, respectively, a helium nucleus, an electron, and a high-energy photon. If the decay involves ejection of a charged particle, the overall charge of the nucleus has changed, and it has become a different element.

The average amount of time that passes before a nucleus decays can range from trillionths of a second to billions of years. When the elements are first created—during a supernova explosion for elements heavier than iron, for example—both stable and unstable combinations are formed. But the most unstable ones decay most quickly, so that as more time passes, only the more stable elements remain. Since the Earth is billions of years old, we would typically expect to find only those radioactive elements that have lifetimes of billions of years or more. For instance, uranium 238 has a half-life of about 4.5 billion years, and potassium 40 a half-life of 1.3 billion years; these are quite useful for geological dating, because we can compare the amounts of these elements present in a sample to the amounts of their stable end products, lead 206 and argon 40, respectively. (Uranium is defined by its 92 protons; uranium 238 has 238 protons and neutrons, and so has $238 - 92 = 146$ neutrons. Forms of elements with different numbers of neutrons are called isotopes.)

How is it then that carbon 14, which has a half-life of only 5,730 years, is useful in archaeological dating? It should all have disappeared a long time ago, leaving only infinitesimal trace amounts in present-day organic matter. The answer is: cosmic rays. The Earth is constantly bombarded by high-energy particles originating from outside our solar system, and these particles produce large showers of lower-energy particles in the upper atmosphere, when they hit the first traces of air at high altitudes. Among the particles created in these showers are neutrons, which then can hit nuclei of nitrogen, which is the most abundant element in our atmosphere. As it happens, when a nitrogen nucleus absorbs a neutron, it very quickly ejects a proton, and the result is a nucleus of carbon 14. (Most of the Earth's carbon is the lighter isotope, carbon 12. Since an isotope is a nucleus with the same number of protons, it acts the same chemically, but since it has a different number of neutrons, the total atomic weight is different.) The new carbon atom quickly binds with oxygen to form carbon dioxide, which circulates readily throughout our atmosphere, and eventually gets incorporated into living matter.

While an organism is alive, it continues to incorporate carbon 14 along with the more common and stable carbon 12, so that the ratio of the two isotopes reflects the prevailing conditions in the environment. Once the organism dies, it ceases to incorporate additional carbon, and the decaying carbon 14 in the object is no longer replenished. The ratio of carbon 14 to carbon 12 starts to drop, decreasing by half every 5,730 years. A measurement of the relative amounts of the two carbon isotopes can then tell us how long it has been since the incorporation of carbon ceased, as long as the remaining amount of carbon 14 is sufficient to make the measurement. This technique is not very helpful for geologists, who mark time in billions of years, but it has proven extremely useful to archaeologists and historians, since it can accurately establish dates back to about 75,000 years before the present.

Other isotope ratios, such as those of molecules found in ice cores, have provided climate data back through several hundred thousand years, from the composition of the water in the ice and also from gas bubbles trapped in the ice. We can reconstruct the temperature from the ratio of deuterium to ordinary hydrogen in the ice, and also from the ratio of a heavy isotope of oxygen to ordinary oxygen. Direct measurement of the gas in the bubbles gives the carbon dioxide concentration, and this percentage is found to correlate extremely well with the temperature, going back several hundred thousand years.

The Faint Young Sun Problem

Paradoxically, one of the problems we have with studying the Sun is that we have so much data that we can develop accurate models. As we saw in Chapter 2, the model is so reliable that we know there is a neutrino problem, in the sense that we are not detecting as many neutrinos as ought to be there. Similarly, the model predicts that the early Sun, billions of years ago, was much fainter than today's Sun.

The reason is this: the rate of nuclear burning depends on how closely pressed together the protons and other nuclei are, particularly down in the core of the Sun where the temperature and pressure are greatest. The rate of burning is *extremely* sensitive to this temperature, since it is mainly controlled by a quantum-mechanical effect (tunneling) that depends strongly on the separation between the particles. As the hydrogen (atomic weight = 1) of the Sun is slowly converted into helium (atomic weight = 4), the average density of the Sun increases a bit, and the core temperature goes up slightly. The higher core temperature produces a higher nuclear reaction rate, which means that more energy per second is being released and consequently the solar luminosity goes up.

The solar models predict that the early Sun was about 30 per cent fainter than today's Sun. Why is this a problem? Carl Sagan was one of the first to point out that the lower solar luminosity means that, for the first 2 billion years of the Earth's history, the average temperature of the Earth was well below the freezing point of water. However, this cannot be correct because sedimentary rocks going back to an age of 1 billion years have been found. There is even some evidence that the young Earth was *warmer* than today. This discrepancy has been given the name "the faint young Sun problem."

The preferred solution to the problem is a large greenhouse effect. As we've said earlier, the Earth's atmosphere acts like a blanket to retain some of the incoming solar energy, so that the Earth is warmer than it would be if it had no atmosphere. The efficiency of this blanketing depends on the composition of the atmosphere, and there are many gases that can make the greenhouse effect stronger, such as carbon dioxide, water vapor, and methane. Sagan proposed ammonia, a very efficient greenhouse gas, as the agent, and this remains one of the possible answers. However, most climatologists think that it was high concentrations of carbon dioxide that were responsible, since early volcanic activity would have released enormous amounts of CO_2 into the atmosphere, and we know that CO_2 causes global warming.

So the solution to the problem is still being debated. No matter what the answer, it is clear that the interaction between the Sun and Earth is quite complex and changeable—yet one more reason why we should not take for granted the present favorable climate on this planet.

Ice Ages

There is good evidence that the Earth's climate varies on time-scales of some tens of thousands of years, mainly due to the solar influence. Variations in the Earth's orbit lead to ice ages via changes in the amount of sunlight falling on different parts of the Earth's surface. (There are also shorter-term [100 to 1,000 year] climate changes that seem to be due to extended periods of higher or lower than normal solar activity levels.)

It is worth noting that humans have never experienced the "normal" climate of this planet. We are now in one of the ice eras, when long-term polar cap glaciation is present, and this seems to be connected with the periodic recurrence of ice ages. Such times are rare on geological time-scales, occurring roughly every half a billion years and lasting for about 100 million years. The Earth is, at most times, considerably warmer than at present and without glaciers at the poles.

On shorter time-scales of one hundred thousand years or less, there have been periodic ice ages during the present ice era, the most recent of which ended about 10,000 years ago. We are now in one of the warmer "interglacial" times, heading for another ice age in a few thousand years. As it turns out, we may have the good fortune to find ourselves in an unusually long interglacial period, and it may be several tens of thousands of years before the depth of the next ice age. This will depend on the correctness of the Milankovitch model, discussed later.

The Debate over the Existence of Ice Ages

Science is a highly dialectical activity, especially in those areas where progress is being made. Where new data are involved, or a new theory is proposed, it is rarely evident immediately whether the new results are to be believed, or whether the new theory is an improvement over the old one. Those proposing change do their part by pushing hard for the new view, and those resisting change are doing their part by arguing that we have a perfectly fine explanation and don't need a new one. Without the former, there would never be any progress; without the latter, there would never be any stability.

This process is well illustrated by the nineteenth-century debate over the existence of ice ages. At the start of the century, the Earth was thought to be about 6,000 years old and the evidence pointed to its origin in a great catastrophe consistent with the biblical accounts of a great flood. By the end of the century it was clear that the Earth is at least several million years old and that there have been repeated periods of extensive worldwide glaciation; and the beginnings of an explanation were being found. The story is filled with ups and downs, unusual characters, disheartening failures and dramatic successes, and only in the past few decades has a generally agreed-upon explanation emerged.

A key turning point was a lecture given in 1837 by Louis Agassiz, a young Swiss scientist, who adopted an explanation he had heard about for the presence of "erratics"—large boulders found many miles from their obvious places of origin—and the numerous gouges and scratches seen on rocks high in the Swiss Alps. One commonly accepted explanation was a flood similar to that described in the Bible. When it seemed difficult to explain the movements of boulders weighing many tons by mere water, a variant of this model involving ice rafts and boulder-laden icebergs was proposed. This was one of the more popular views at the time of Agassiz's lecture, although it carried the disturbing implication that the water worldwide had reached a height of at least 5,000 feet, since erratics were found high in the mountains. Agassiz argued instead that a massive sheet of ice had covered the region, moving boulders, carving the gouges, and leaving behind massive deposits of silt as they retreated.

It took some 30 years for the view that widespread glaciation had occurred in the fairly recent past to become accepted. Examination of glacial activity in such places as the Alps, the far northern United Kingdom, and the American northeast, provided convincing evidence via direct examples of the processes that needed to be explained. Glaciers can reach a maximum thickness of about 1 mile before the pressure of the thick layer causes the ice at the bottom to partially liquefy and begin flowing. (This process also applies to mountains, which can therefore not be more than about 10 miles high under Earth's gravity.) As more snow falls on top of the glacier, an equal amount of dirt and rock-laden ice is forced out of the bottom and edges. During warmer times, the edges retreat, leaving behind a characteristic type of deposit. Geologists have been able to identify the type of deposit left by each of the many processes involved and to identify the history of the advances and retreats of glaciers, eventually constructing a global map of their history since the last ice age.

When Agassiz was persuaded by John Lowell to accept a position at Harvard in 1847, he found that American geologists were enthusiastic about the new theory and were able to explain the sediments and polished rock surfaces prevalent in, for instance, western New York state, by glaciation. On both sides of the Atlantic, geologists were finding that the new theory helped to explain observations that had long been puzzling. A prototypical example of the way a scientist experiences a new paradigm is provided by the conversion of the great British geologist Charles Lyell to the glacier theory: "On my showing him a beautiful cluster of moraines within two miles of his father's house, he instantly accepted it [the glacial theory], as solving a host of difficulties which have all his life embarrassed him."[2] By the end of the nineteenth century the view that the Earth has experienced periodic ice ages was accepted. The problem then became one of explaining why these major climate changes occur.

The Milankovitch Cycle

The way in which science typically proceeds is by guessing at a theoretical explanation—what Einstein called a "free creation of the human mind"—and testing the consequences of the theory against what is actually observed. Theories are not proven; they are, at best, not disproven. The philosopher Karl Popper summarized this process as "conjectures and refutations" and proposed that a theory can only be called "scientific" if it makes predictions that can, in principle, be falsified. Put another way, a theory that cannot in principle be disproven (Popper's example was psychoanalysis) cannot be called scientific.

However, it is often easier to propose theories than to test them, so that the problem generally is one of ruling out the wrong theories, and hoping that a plausible one remains. Dozens of theories were proposed in the nineteenth and early twentieth centuries to explain ice ages, most of which were shown to be inconsistent with the evidence, or were untestable and therefore left in a limbo state. It took a full century from the date of Agassiz's famous lecture before a convincing theory was formulated to explain the recurrent glaciations.

The first step toward the modern explanation of Ice Ages was taken very early on. In 1842 the French mathematician Jacques Adhémar

2. From an 1840 letter to Agassiz written by the Rev. William Buckland, geology professor at Oxford. Quoted in John Imbrie and Katherine Palmer Imbrie's *Ice Ages: Solving the Mystery*.

proposed that the tilt of the Earth's orbit combines with the elliptical shape of the orbit to produce an imbalance in heating that leads to glaciation. These days the Northern Hemisphere is closer to the Sun in winter than in summer, but the Southern Hemisphere is farther away during its winter. More subtly, the Earth moves more slowly when it is farther from the Sun and more quickly when it is closer. The result is that winters last longer in the Southern Hemisphere at the present time, and winter is when nights are long and days are short. The upshot is that there are more hours of darkness than of daylight in the Southern Hemisphere. That hemisphere must therefore be growing slowly colder owing to the imbalance between heating and cooling.

This model predicts that ice ages will alternate between the Northern and Southern Hemispheres, and that the cycle time between north and south will equal the period of precession of the Earth's tilt axis, or 21,000 years.[3] The Northern Hemisphere would have been glaciated half a cycle in the past, about 11,000 years ago, which is indeed about when the last ice age ended. However, it turns out that Adhémar's reasoning was faulty, and that the total solar energy at either pole—the total number of calories it receives over an entire year—does not show the imbalance needed in this theory: any decrease during one season is exactly balanced by a compensating increase during the other season. The total amount of heat received by one pole over the course of a year is the same as the total received by the other pole.

The next major step toward an astronomical theory of ice ages was taken in 1864 by the self-educated Scottish geologist James Croll. The son of a stonemason, and too poor to afford to go to university, Croll failed successively as a millwright, a carpenter, a teashop owner, and a hotelkeeper, ending up as a janitor at a small college in Glasgow. The

3. Those familiar with astronomy might ask why we say that the precession period is 21,000 years, when it is well known to be 26,000 years. The explanation is that we are speaking about two different precession periods, one relative to the "fixed" stars and the other relative to the major axis of the Earth's orbit. The first, with a period of 26,000 years, is the rate of precession of the Earth's seasons relative to the stars. The latter is the one that is relevant to the issue of solar influence on Ice Ages; in round numbers it is 21,000; in fact, the precession of the axis is quasi-periodic, with periods of 19,000 and 23,000 years.

college had a good scientific library and Croll had ample time to think about the glaciation debate. He came across Adhémar's theory and quickly realized that it was wrong, but he now added an additional factor: the amount of eccentricity of the Earth's orbit changes with time, so that the winter-summer difference is also variable, both with a period of 100,000 years. Croll reasoned that, even though the heating averaged over a year is balanced, there are epochs lasting tens of thousands of years with long, cold winters. His idea was that such seasons would cause large amounts of snow to accumulate, and he also argued that the additional snow cover would reflect back more of the incoming sunlight, so that the cooling effect would be amplified. This effect is called positive feedback, and it is indeed a prominent feature of all modern climate models.

Croll went further, and realized that the patterns of air circulation, the wind patterns shown in Figure 7.4, would drive corresponding ocean currents that also transport heat from the equator to the poles. He proposed that an increase in the equator-to-pole temperature difference would increase the strengths of the winds and cause them to shift equatorward, and that this would cause the strong westward current in the Atlantic to hit Brazil at lower latitudes, which would deflect the Gulf Stream. This astounding bit of reasoning is correct, and it revealed another feedback mechanism that can also amplify the effect of a small change in solar heating on the global climate.

Croll's theory was one of the few at the end of the nineteenth century that fit Popper's criterion of falsifiability, since it made predictions that could be tested against the geological record. Unfortunately, despite some successes, such as the discovery that there had been periodic ice ages, it eventually failed this test. The theory predicts that ice ages will alternate between the Northern and Southern Hemispheres, and that glacial epochs coincide with times of large orbital eccentricity, the last of which ended 80,000 years ago. This contradicts the geological evidence that the most recent ice age ended about 10,000 years ago. By the beginning of the twentieth century Croll's theory was nearly forgotten.

The astronomical theory was developed in its modern form by a Serbian engineer named Milutin Milankovitch. As a young academic in applied mathematics in Belgrade, he decided to use his talents to solve some grand problem, and he settled on developing a mathematical theory of climate—not just for the Earth but also for Mars and Venus, and not just for the present, but for all past and future times. It took him 30 years, but he succeeded.

Milankovitch approached the problem by first calculating the planetary orbits, including eccentricity, tilt, and precession. From this he began the laborious process of calculating the insolation (the rate of incoming solar energy) for each planet throughout the year and at different latitudes on the planet's surface. As he himself noted, had he been younger he would not have known enough to do the calculations, and had he been older he would have known better than to try.

His early publications, written in Serbian and published during the First Balkan War, went unnoticed. In 1914, during World War I, Milankovitch was captured and put in prison, but a Hungarian mathematician heard about his plight and had him released and sent to Budapest, where he worked for the next 4 years. In 1920 he published his first major result, a mathematical theory for the present-day climates of the Earth, Mars, and Venus, which happened to catch the attention of a German climatologist named Wladimir Peter Köppen and his son-in-law Alfred Wegener. The latter is remembered mainly for his theory of continental drift, which, like Milankovitch's own theory, would fall into disfavor and then return triumphantly many years later.

Working together, they agreed that the major change needed in Croll's theory was that it is cool summers, rather than cold winters, which lead to ice ages. Since winters are cold enough anyway, ice is going to accumulate, but the issue is whether or not it will melt away in the summer (averaged over many, many years, of course). If there is not more melting than accumulation, the glaciers will expand, leading to an ice age. The theory Milankovitch developed is now generally accepted,

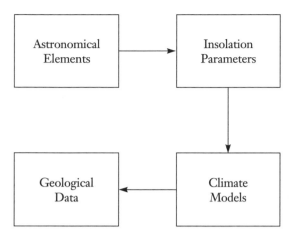

FIGURE 7.6   The four major steps in testing an astronomical theory of long-term solar influence on climate.

although not without a 50-year struggle involving detailed geological evidence and a protracted debate over the interpretation of data from the half-million-year record found in deep ice cores (Figure 7.6). The advances and retreats of glaciers, which are a measure of the changes in global climate over the past half a million years, are found to have a periodic behavior that can be linked to the periodicities in the Earth's orbital parameters: 19,000 and 21,000 years, due to the quasi-periodic precession of the Earth's tilt axis; 43,000 years, due to the oscillation in the size of the angle of tilt of the axis (which equals the obliquity); and 100,000 years, due to the oscillation in the magnitude of the eccentricity of the Earth's orbit. These three types of orbital changes are illustrated in Figure 7.7.

GLOBAL WARMING

There is now no doubt that the Earth has been warming during the entire twentieth century. Major studies by the Intergovernmental Panel on Climate Change (IPCC) published in 1990, 1995, and 2000 have carefully examined over 100 years of measurements and concluded that

1. Precession of tilted
   rotation axis.

2. Variation in tilt angle

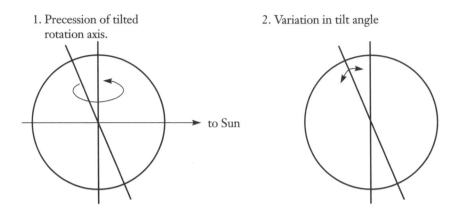

3. Variation in ellipticity of orbit

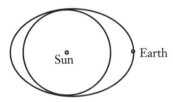

FIGURE 7.7   The three types of variation in the Earth's orbit relative to the Sun, which according to Milankovitch's theory, are thought to cause periodic ice ages.

the average global temperature has risen nearly 1°C (roughly 2°F). This may seem to be an insignificant amount, until one notices that average sea levels have risen about 6 inches during that time, that alpine glaciers have retreated several miles, and that the average thickness of Arctic ice has decreased from 10 feet in 1976 to 6 feet today. Expeditions to the North Pole in the summer of 2000 found thinner than usual ice all along and even a mile-wide patch of open water at the pole itself. Though holes in the polar ice often occur, the thinning ice and the polar water dramatized global warming significantly. The global average temperature during the depths of the last ice age was only about 4–8°C (7–14°F) lower than today's, although in some places such as North America it was quite a bit colder with strong winds. In recent years, global temperatures have continued to rise sharply. The year

1998 appears to have been the warmest on record—even after we remove the influence of El Niño—with 1999 close behind, and the winter of 2000 was the warmest ever recorded. The outbreak of tropical diseases, such as malaria and mosquito-borne encephalitis, in northern latitudes is also being attributed to higher global temperatures and the associated changes in rainfall patterns. The spread of the mosquito-borne West Nile virus in the northeastern United States is also fostered by these changes. The rise in temperature during this century is therefore significant, and further increases could have major worldwide consequences.

It is also the case that emissions of greenhouse gases, such as CO_2, have increased dramatically during this century. While it is tempting to blame the warming on these man-made pollutants, there are some difficulties in establishing a direct connection. For instance, CO_2 levels have risen steadily during the century, but global temperatures rose rapidly in the first half of the century, then leveled out for two decades, then started rising again. Perhaps some storage of energy was involved (the deep ocean temperatures rose during this time), but at a minimum we know that the connection is not simple and straightforward. Sophisticated climate models are needed in order to determine how much of the temperature rise is natural and how much is man-made, how much is immediate and how much is delayed; and models are only now becoming good enough to make such determinations.

We need to factor into this discussion the effects of solar variability on time-scales of the order of 1,000 years. There is evidence for this connection from indicators such as sunspot and radioactive carbon 14 records, but there is also evidence for a change in the relative dominance of solar effects on global warming since the start of the Industrial Age, meaning that the Sun used to be the main controller of climate change, but that man-made effects are now beginning to dominate. In the following discussion, we will examine human effects on global climate, on about 100-year time-scales. Evidence for global warming and for the relative contributions of solar variations and of man-made effects are also discussed. We conclude by considering the probable future course of events.

The Medieval Grand Maximum and the Little Ice Age

There is a famous study of the relationship between the Sun and the Earth's climate that reminds us to be cautious about taking statistical correlations to be meaningful. In 1923 a British climatologist published the results of a study comparing the sunspot number to the water level of Lake Victoria in Africa. The data covered two full solar cycles, and a comparison of the two records is quite convincing: anyone looking at the two curves would agree that there is a statistically significant correlation between them.

However, no sooner had the paper appeared than the correlation between sunspot number and water level in Lake Victoria completely disappeared. Since 1923 there has been no correlation whatsoever between the two. We are left wondering whether there was a real effect that became masked by some other climate change, or whether it was all merely a statistical accident in the first place. Sadly, we have no way of knowing, at least for now.

With that caution in mind, we will describe some of the apparent connections between changes in the Sun and changes in the global climate on the Earth.

One of the more productive figures in sun-climate research of the century was Charles Greeley Abbot. He lived to the age of 101 (1872–1973) and published results over seven solar cycles. He was a powerful figure in American science, and served as the Secretary (the head) of the Smithsonian Institution in Washington until he retired in 1944. His main work, starting in 1902, only a few years after he joined the Smithsonian in 1895, was in measuring the solar constant, which he did with bolometers, instruments for measuring the total radiative power integrated over all wavelengths. By measuring the total energy emitted from the Sun year after year, he hoped to determine whether the Sun's output was variable. The main limitation to the accuracy of his measurements was in finding the variable contribution of the Earth's atmosphere to his measurements, and with hindsight we can see that he never overcame this problem, in spite of numerous efforts such as the setting up of remote stations to make observations through different bits of atmosphere in the hope of removing local effects. Abbot was convinced that he had found variations in the solar constant of 3 per cent to 10 per cent and that he had correlated them with the Earth's climate, a conclusion with which scientists today do not agree, because more accurate measurements from space show the effect to be about a tenth as large as Abbot thought. A restudy of the long run of data carried out in the 1970s also detected such smaller variations, but Abbot's conclusions about a solar-terrestrial link were not correct.

Yet the climate record does show evidence of an 11-year influence, which would imply that the Sun is somehow driving the changes. One recent proposal involves cosmic ray influence on cloud formation, with the energetic particles acting as seeds for droplet nucleation in the atmosphere. The availability of more powerful computers may play an important role in answering the question, because more realistic global climate models can now be run. This is clear in a recent paper from the Goddard Institute for Space Studies, which argues that the varying levels of ultraviolet radiation from the Sun can change the atmospheric circulation patterns high in the atmosphere, which then produces a change in the circulation even at ground level. The overall global temperature is only slightly altered, but the local climate at a given location can change significantly.

There have also been changes in climate over the past few centuries, on time-scales longer than the 11-year solar cycle but much shorter than the tens of thousands of years associated with changes in the Earth's orbit. Figure 7.8 gives a rough indication of how global temperatures have varied during the past 1000 years. There are two important features to notice: during the so-called High Middle Ages, the climate was noticeably warmer than it is today, and starting about 1300 until the 1800s the climate was quite a bit cooler. These periods are often

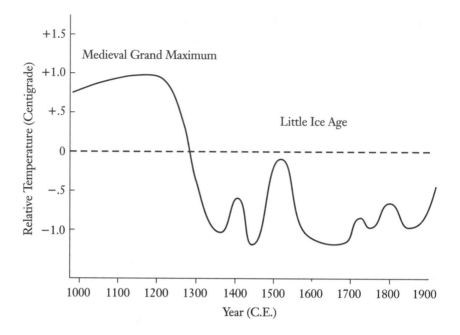

FIGURE 7.8   Schematic illustration of the global temperature variations during the High Middle Ages into modern times.

called, respectively, the Medieval Grand Maximum and the Little Ice Age. It seems that these little ice ages come and go about every 2,500 years, and there is no known cause for them. Variations in the solar activity level have been postulated, but evidence for this is scanty.

If we take the period to be 2,500 years, then we are presently on the upward slope and will continue to be for another 1,000 years. This will be followed by another cooling period and will probably add to the general downward trend driven by the astronomical effects. The Milankovitch model predicts that the depths of the next true ice age will occur about 23,000 years from now.

The Industrial Age

Merely measuring the total energy coming from the Sun may not be enough to understand the ways in which the Earth's climate is affected by solar variability. For example, it has been suggested that cosmic rays—those energetic particles that are always bombarding the Earth—might play a role in cloud formation via a "seeding" effect as they deposit energy in the upper atmosphere. It is known that the activity cycle of the Sun changes the number of cosmic rays that hit the Earth, and this could be how the solar cycle shows up in our weather. Another possibility is via the short-wavelength ultraviolet radiation from the Sun. This too varies with the solar cycle, and we know that this radiation is absorbed high in the Earth's atmosphere, causing the formation of the ozone layer. The amount of UV varies with the solar cycle, so that the amount of energy deposited in the atmosphere varies as well, and the result is a change in the temperature of that portion of the atmosphere. This temperature change causes a change in the global circulation pattern of the winds, which then changes the weather map. Such is the proposed mechanism, at least.

There is, therefore, at least the possibility that small variations in the radiation and energy coming from the Sun can have a measurable influence on the Earth's climate. In order to test this idea, we need good measurements of the Sun's output and we need good measurements of the Earth's climate. Such data are available now, but were not available in the past. Reliable temperature data go back about 150 years, and reliable solar data go back only 20 years. To push the solar data further back, we use indirect methods, such as the correlation between sunspot number and the solar radiation output; those data go back to about 1610. For the Earth's climate in earlier centuries we can use tree rings and pollen, as well as precise measurements of certain isotope ratios in ice cores, and the organic composition of sea corals.

Any method we use becomes less reliable the further back in time we look, but we are now fairly certain that we have good determinations for the past 1,000 years. The result is striking: up to the nineteenth century, the changes in the Sun and the changes on the Earth correlate very well, meaning that changes in the Earth's climate could have been due largely to solar changes. In the nineteenth century there were some major volcanic eruptions that cooled the Earth for several years by throwing an obscuring veil of dust high into the upper atmosphere.

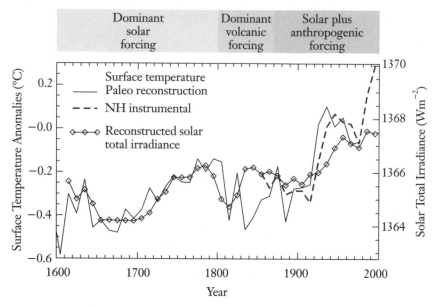

FIGURE 7.9   Comparison of solar irradiance with temperature at the Earth, from 1600 to the present. Significant volcanic events kept global temperatures low in the nineteenth century, but the Northern Hemisphere temperature ("NH") has risen more steeply than has the solar input for most of the twentieth century. It has risen especially quickly in the past few years.

The correlation in those years is straightforward between the global climate and these rare and unusually large terrestrial events. In the twentieth century—that is, since the Industrial Age—there have not been any such enormous volcanic eruptions, but the correlation with changes in the Sun has nevertheless broken down. It seems that the changes in the Earth's climate can no longer be attributed to natural effects from the Sun, but may be due mainly to human activity ("anthropogenic forcing"), such as increased carbon emissions into the atmosphere (Figure 7.9).

Implications

There is very little doubt that the increased emission of industrial pollutants such as CO_2 is causing global warming, which will continue as long as the CO_2 levels continue to increase. But we have seen that it is not at all obvious that the consequences of human activity have shown themselves fully in the global climate system. The linkage of chemical and biological processes involved in the worldwide "carbon cycle" is extremely complex, involving the atmosphere, the oceans, plant and animal life, and the soil and sediments. Each of the last three can add carbon to or remove it from the atmosphere, but our best estimates indicate that between 1950 and 2050 atmospheric CO_2 will double.

What would be the consequences of a doubling of CO_2 in the atmosphere? The climate models predict that global temperatures would increase an average of about 6–10°F (3–5°C). This may not seem to be much of an increase, until we realize that the global average temperature at the depth of the last ice age 21,000 years ago was "only" 7°F (4°C) lower than today's average. The average during the Little Ice Age from the fourteenth to the nineteenth centuries was only 1°C lower than today's, and the warming at the peak of the present interglacial, about 4,000 to 8,000 B.C.E., was only about 1°C above today's average. A change of a few degrees is therefore large and significant in terms of its effect on the global climate.

Equally important, higher global temperature means increased evaporation, which leads to increased precipitation; the number of storms increases when the oceans are warmer. More to the point, higher ocean temperatures mean larger, more powerful storms. We had a small taste of this effect in 1997–98 with the El Niño–Southern Oscillation. Global warming therefore means an increase in the number of devastating natural disasters. As if this were not enough, water vapor is itself a greenhouse gas, so that a rise in global temperature leading to increased evaporation can itself lead to a further rise in global temperature, which causes even more evaporation, and so on. This is known as a runaway greenhouse effect.

It might be tempting to argue that, since the world is now undergoing a gradual decline in temperature based on the Milankovitch theory of ice ages, the man-made warming may prevent us from descending into another ice age. But there are several problems with this reasoning. First, the time-scales involved are very different: the next ice age is coming, but it is thousands of years away, whereas the global warming due to fossil fuel burning is arriving very quickly, within a few decades. Human activity might then cause an enormous upswing in global temperature followed by a more drastic downturn than would otherwise have occurred. Moreover, the warming that is now either under way or imminent has not been intentional, but rather is a side effect of our efforts to produce energy for an industrial economy. Given our present rudimentary understanding of global climate, it would be difficult to produce a controlled, planned change. The likelihood that an unplanned, uncontrolled change would be beneficial is extremely low.

In the United States, despite some recent arguments, the question of whether global warming is a real problem was settled in 1992 when the Bush administration and Congress ratified the United Nations Framework Convention on Climate Change. The details of how the goal of reducing greenhouse gases in the atmosphere to "a level that would prevent dangerous anthropogenic interference with the climate system" were left to a later meeting, which was held in Kyoto at the end of 1997. That meeting was inconclusive, largely because developing countries did not agree to reduce emission standards. Neither the United Nations "climate summit" held in Buenos Aires in 1998 nor the meeting in The Hague in 2000 reached consensus, although the United States signed the protocol initiated in Kyoto. It is already clear, unfortunately, that the goals of the Kyoto protocol are not being met. Ratification of the protocol by enough developed nations to put it in force is not in sight.

Individuals can take steps to help in reducing greenhouse gas emissions, independent of any governmental actions. These are rather prosaic steps, and many of them became commonplace in the late 1970s; they are, however, effective, and could achieve more than half of the reduction that was targeted for the United States at the Kyoto meeting:

- Winterize your home. If every American household wrapped its hot water heater in an insulating blanket, replaced inefficient showerheads, caulked doors and window frames, used doublepane glass in windows, and insulated its attic, U.S. carbon emissions would be lowered by more than 100 million tons a year.

- Drive a fuel-efficient car. Between 1973, the year of the first "gas crisis," and 1988, the fuel efficiency of the average U.S. automobile doubled; since then, the average has not changed. Had the earlier trend continued, U.S. emissions would now be about 150 million tons a year lower. Low gas prices during the decade of the 1990s contributed to the problem by encouraging the purchase of larger, less fuel-efficient cars, vans, and trucks. The new types of high-efficiency hybrid cars consume far less gasoline per mile and seem to be an attractive option for reducing carbon emissions. Still, low-gas-mileage sport utility vehicles (SUVs) are very popular.

Worldwide, the emission of carbon into the atmosphere is being contained a bit by the movement from coal to oil to natural gas, which represents a trend toward lower carbon per unit of energy produced. Moving to hydrogen as a fuel entirely eliminates carbon emission, since it does not contain any carbon.

There are some measures for reducing, or at least slowing, the rate of carbon emissions that will require government action for their implementation. Some examples would be tax incentives for installation of energy-efficient machinery and a tax (whether or not it is so named) on carbon emission. In one form or another, there is a cost associated with the production of pollutants, even if it is not obvious at the moment. One approach would be to ignore the problem and pay later, which would probably require costly and difficult technological solutions. Another approach is to prevent the problem in the first place by reducing carbon emissions, which requires a present-day cost, but would certainly be less painful in the long run. The possibility of trading emissions in one place for reductions elsewhere is a short-term solution discussed by economists and was a major debating point in The Hague in 2000. In the end, carbon dioxide put into the atmosphere anywhere spreads around the world.

Space Weather

In several of the earlier chapters in this book we have mentioned how safe and sheltered we are, living on the surface of the Earth beneath a protective blanket of atmosphere and magnetic field. Conversely, when we venture outside of our safe haven, we step into hazardous territory. "Empty" space is actually far from empty: it is filled with high-energy particles and radiation, bullets of matter shooting in all directions, clouds of hot plasma thrown out by the Sun, and extremes of heat and cold (simultaneously).

The Sun is one of the major sources of energetic particles and radiation that affect the Earth. The solar corona, even when it is not producing a major eruption, is so hot that the enormous gravitational pull of the Sun cannot completely contain it. The result is an expansion of its million-degree plasma into interplanetary space to form the solar wind, that roars outward at supersonic speeds of hundreds of miles per second. Rapid, intense explosions in coronal active regions, the solar flares, produce bursts of x-rays, gamma rays, and high-energy particles at the Earth. Large-scale reconfigurations of the magnetic field of the Sun cause coronal mass ejections that send enormous clouds of hot ionized plasma and magnetic fields toward the Earth, often with severe consequences.

Our small planet glides innocently through these perils at about 19 miles per second in its yearly orbit around the Sun. Its thin atmosphere provides some protection from the particles, radiation, and debris, and its weak magnetic field puts a shieldlike barrier out into space which deflects all but the most energetic charged particles. Inevitably, though, the portions of our upper atmosphere that absorb the radiation are affected by it, and the Earth's magnetic field is disturbed and distorted by its interaction with the solar wind and the mass ejections. This kind of dynamic activity in Earth's vicinity produces a host of effects that are now generically referred to as space weather.

In the same way that weather forecasts are issued by the National Weather Service (NWS), the Space Environment Center (SEC) in Boulder, Colorado, issues space weather alerts, watches, and warnings. (Both organizations, NWS and SEC, are now housed in the same office building, part of the National Oceanographic and Atmospheric Administration, NOAA.) The SEC defines these three levels of announcement as:

1. Watch: A watch is issued when conditions are favorable for a certain level of activity to occur.

2. Warning: A warning is a high-confidence prediction of imminent activity.

3. Alert: An alert provides rapid notification to recipients that an activity-detection threshold has been reached.

The function of the SEC is "providing space weather alerts and warnings to the nation and the world for disturbances that can affect people and equipment working in space and on Earth." This information is available at the SEC Website, http://sec.noaa.gov. In this chapter we examine the solar conditions that cause these disturbances.

THE SOLAR WIND

By the middle of the twentieth century, it was becoming clear to solar physicists that there is more than light and heat coming from the Sun and influencing the Earth. Both strong auroras and disturbances of the Earth's magnetic field were seen to have a recurrence period of 27 days, which is the same as the rotation period of the Sun as seen from Earth. It was also known that auroras and magnetic disturbances are stronger near the Earth's poles, which indicates that whatever is coming from the Sun is being funneled up into the polar regions. The obvious assumption—for a physicist—is that charged particles, electrons, protons, and ions are being deflected toward the poles by the Earth's magnetic field, and that they are being spewed out from some location on the Sun. Such a suggestion had been made as early as 1908 by Kristian Birkeland to explain the aurora. This emitting region on the Sun apparently lasts for at least a few solar rotations. Although a name was given to these hypothetical areas—M regions—by Julius Bartels in 1932, the source of the magnetic storms could not be found on the Sun. Several different lines of work had to come together.

During the early 1950s, the German theoretical physicist Ludwig Biermann published a series of papers on comet tails, as we saw in Chapter 6. He argued that the usual explanation for why comet tails always point away from the Sun—radiation pressure, from the sunlight falling on the comet tail particles—was grossly inadequate to explain the observations. Instead, he proposed that a stream of high-speed charged particles flowed out from the Sun. He even calculated the speed, 300–600 miles per second, and the number of particles in the flow, roughly 1,000 ions and electrons per cubic inch.

Not everyone agreed, and the theory was not accepted for many years. Meanwhile, a mathematician turned solar physicist named Sydney Chapman decided to calculate how far out from the solar surface the corona should extend. His inspiration for doing this calculation was the realization that a hot plasma has many free electrons in it, and they are extremely efficient at carrying heat. This means that energy from a hypothetical coronal "base" near the surface of the Sun can be transported rapidly outward into the upper parts of the corona. The result is that, instead of the temperature dropping rapidly as one proceeds outward, it falls very slowly. For instance, if the coronal temperature is

1 million degrees very close to the Sun, then it is still 200,000 degrees at the distance of the Earth. The corona is greatly extended, although sparsely populated at 1 A.U. (a unit equal to the average distance between the Earth and the Sun, 8.3 light minutes), and we might even say that the Earth is in the corona.

Chapman's calculation was done for a static corona, in which the gas extends upward to some height because of its temperature, but which basically is held in place by gravity. In 1957, as he was about to take a position at the University of Chicago, the physicist Eugene Parker began working on the problem of explaining the solar cycle–related variation in cosmic ray intensity. Cosmic rays fill interstellar space and constantly bombard the Earth. But the number of cosmic rays varies with the solar cycle, being lowest near solar maximum, particularly for the weakest (lowest-energy) particles. It could be that a high rate of particle and magnetic field outflow from the Sun during years of high activity helps to push back the incoming cosmic rays, so that fewer of them reach Earth when the Sun is active. Parker discussed the possible role of the Sun with both Chapman and Biermann, and soon realized that he had to find a way to reconcile Chapman's static but extended corona with Biermann's outflow of material from the Sun.

The key turned out to be in the far reaches of the solar system, out beyond the orbits of Jupiter and Saturn, where the solar system merges into the space between stars, the so-called interstellar medium. This region has a very low density of particles, and Parker noted that Chapman's model did not have the corona diminishing quickly enough: Chapman's model predicted a fairly large gas pressure at the junction with the interstellar medium, far higher than what is actually out there. (The words "large" and "small" are relative terms here. Parker calculated the pressure to be 10^{-24} [a millionth of a millionth of a millionth of a millionth] pound per square inch, but Chapman had calculated it to be more than 10,000 times greater.) Parker considered the possibility that Biermann's outflow of material was due to a continual expansion outward of Chapman's extended corona.

By solving the relevant equations, he found that a unique solution for an outflowing solar wind existed that was physically realistic (in particular, that started at the surface of the Sun with a low, rather than a high, velocity) and that also had a low pressure at infinity (which is the technical term for "very far away"). This solution, for a coronal temperature of a few million degrees near the Sun yields an outflow speed at 1 A.U. of about 1,000 miles per second and a particle density of a few thousand ions per cubic inch, just what was required by Biermann's calculations.

The theory was highly controversial and was not well received. It was not until the 1960s, when spacecraft actually measured the particles flowing out from the Sun, that the high-speed solar wind was accepted as real.

THE EARTH'S MAGNETOSPHERE

It has been known since the publication of William Gilbert's *De Magnete* in 1600 that the Earth has a magnetic field, as if it were a giant magnet (the full title of his work is *On the Loadstone and Magnetic Bodies and on the Great Magnet of the Earth*). The work was based on many years of observation and experiment with iron-rich magnetized rocks, or lodestones, and it was motivated in large part by the need to explain to mariners why their compasses worked, and also why they sometimes failed to work. In any case, we know that the Earth is surrounded by a large shell-like invisible magnetic field, whose existence is inferred from its effect on things like magnetized needles.

This field extends far out above the Earth's surface and into space (Figure 8.1). Note that the north magnetic pole of the Earth is in the geographical south these days; there are reversals of the Earth's field about every million years or so. If there were no disturbances from the Sun, the shape of the field would be very simple and not terribly interesting, as shown in the figure. But the Sun does send out a wind of energetic ionized particles. As this wind blows past the Earth, it produces a profound distortion of the Earth's magnetosphere—the magnetic volume surrounding the Earth—as shown in simplified form in

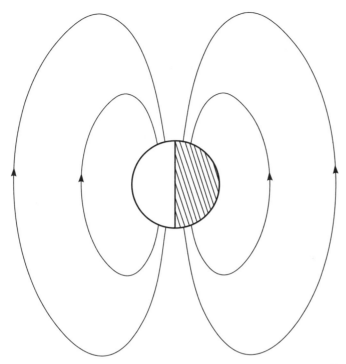

FIGURE 8.1   The Earth's magnetic field without solar wind effects.

Figures 8.1 and 8.2. The dynamic interaction between the solar wind
and the magnetosphere changes the simple, dull, bar-magnet pattern
into one rich in complications and in hazards to man and machine,
both in space and on the ground.

The main effects are:

1. The Earth's magnetosphere deflects the solar wind, which
then follows a complex route around and partly into the vicinity of
Earth.

2. The solar wind carries some of the Sun's magnetic field with it,
which opens up and reconnects with a portion of the Earth's field.

3. A huge current circuit, carrying over 1,000,000 amps, flows from
the magnetosphere, through the upper atmosphere and out again,
with an associated auroral display.

4. A long tail of highly dynamic plasma forms in the anti-Sunward

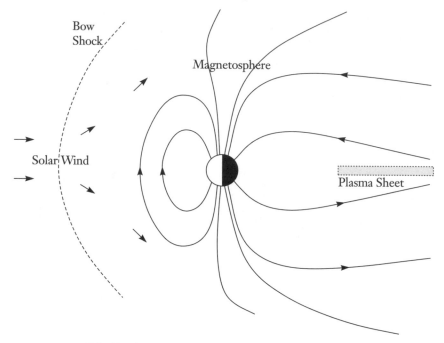

FIGURE 8.2   The Earth's magnetic field disturbed by, and interacting with, the solar wind.

direction, and is subject to instabilities that cause dramatic effects at Earth.

5. Energetic charged particles are trapped in torus-shaped "radiation belts" near the Earth.

Auroral Particles and Global Currents

The auroras, or Northern Lights, are spectacular displays consisting of glowing curtains of light shimmering high above the ground and lasting for many hours or days. It has been known for decades that the Sun somehow causes these displays, and for many years it was thought that energetic particles from the Sun impinge directly down onto the Earth's atmosphere, releasing their energy as they strike the first traces of air. The mechanism responsible for the light show is ionization—the excitation of electrons in an atom to a higher quantum level or their expulsion out of the atom entirely—of oxygen and nitrogen atoms, followed by a recombination of an electron with the ionized atom or the return of an electron from a higher level to a lower level in the atom. These processes lead to the release of energy from the atom, in the form of a photon, and it is this light that we see as the aurora.

We now know that this simple and direct view is not correct, and that auroras are caused by a somewhat more complicated process. The flow of energetic charged particles from the Sun acts as a generator, via the same basic process involved in a man-made generator: moving a conductor through a magnetic field causes a flow of charged particles in the conductor. In this case, the conductor is the high-speed plasma from the Sun blowing past the Earth, and the magnetic field is that of the Earth extending out high above the Earth's surface.

A flow of current in one way or another involves a circuit, with the current flowing along a complete, closed conducting path. In the case of auroras, the path of the circuit includes the upper atmosphere of our planet, a region called the ionosphere that has free electrons that can carry a current. As we mentioned in Chapter 6, the ionosphere was identified early in the twentieth century from its effect on wireless communications, and it has been known for some time that auroras occur more than 60 miles above the Earth's surface, at ionospheric heights. But it is only recently that the flow of currents, in the form of an auroral oval, has been identified (Plate XIV), mainly through imaging from above by satellites. The path taken by the solar wind-induced current is down from the region of interaction between solar wind and magnetosphere at the local dawn, around the auroral oval to the local dusk, and back up and out into the magnetosphere. There are actually two currents, one flowing westward on the day side, and the other flowing eastward on the night side. They are known as the auroral electrojets, and typically carry roughly 1,000,000 amps of current.

A current is a flow of charged particles, typically electrons since they are low-mass and therefore easily moved. The current flow described above, downward from the magnetosphere, around the auroral oval, and back up to the magnetosphere, implies a flow of electrons down, around, and back up. The downward-flowing electrons deposit energy when they collide with the first few atoms of air high up in the atmosphere. These collisions excite and ionize the atoms, and the relaxation of these excited atoms back to their normal state involves the release of photons of light, which we call the aurora.

The downward streams of electrons concentrate into long, thin vertical sheets, so that auroras tend to have the appearance of curtains of light. Their greenish-blue color is due to the excitation of oxygen atoms, while an occasional appearance of a light red color near the bottom of the curtain is due to excitation of nitrogen atoms. The waviness or sideward motions of auroral curtains is not due to an actual motion of the curtain of light, but rather to a change in the place of injection of the down-flowing electrons—similar to the apparent motion of water directed toward a bed of plants from a moving garden hose. A closer analogy would be the electron beam in a TV tube, which is swept left-right and up-down across the phosphor screen to map out the brightness pattern of the desired image.

With satellite observations now available, the auroral oval is seen to be a nearly permanent feature, varying in size and strength as the flow of hot magnetized plasma from the Sun varies. When there is an especially powerful eruption from the Sun (also having a magnetic field that is oriented in such a way that it interacts especially strongly with the Earth's field), there is a violent increase in the ionospheric currents and the associated aurora. Such an outburst is known as a geomagnetic substorm, a rather dull name for an event that generates a million megawatts of power for several hours, 10 times higher than the usual level.

Reconnection

The solar wind flowing past the Earth, carrying solar magnetic field with it, has a profound effect on the magnetosphere. In fact, it is the solar wind that confines the Earth's magnetic field to a cavity around the Earth, which we call the magnetosphere. The magnetized plasma of the solar wind compresses the magnetosphere, and a pressure balance is established between the two systems at a distance of about 10 Earth radii out, in the Sun-facing direction (the upstream side). On the downstream side, opposite to the Sun, the magnetosphere is stretched out into a long tail extending indefinitely far back.

The two magnetic field regions—Earth's magnetosphere and solar wind plasma—are separated at a surface called the magnetopause. The word "pause" is a generic term that can be translated "edge"; for instance, the outer boundary of the solar wind influence in the outer regions of the solar system is called the heliopause. In this first-level picture, the solar wind and its entrained magnetic fields slip around the magnetopause and continue on past the Earth, with very little additional interaction beyond the distortion of the geomagnetic field.

But life is not so simple. The magnetic field emanating from the Sun does not always merely slide past. Under the right circumstances, the solar and terrestrial fields interact violently, via a process called magnetic reconnection.

Figure 8.2 showed some magnetic field lines near the Earth's poles that extended out but did not close back to the corresponding region of the Earth in the other hemisphere. These are "open" rather than "closed" field lines, and they are present because of the magnetized plasma being thrown out by the Sun toward the Earth. Notice that the part of the Earth's magnetosphere that faces the Sun has a field that is oriented down-to-up, that is, from the Earth's geographic south pole up and around to the north pole. This field is normally closed—it starts at one place on the Earth and ends at another place on the Earth.

However, the solar wind brings with it some of the Sun's magnetic field, and that field may be oriented either in the same direction as the Earth's field or in the reverse direction; the latter case is shown in Figure 8.3. If the solar field is directed oppositely to that of the Earth, then an interaction between the two can take place, in which the field lines change their connectivity. The fields can disconnect and then reconnect in a new way, as shown in the figure.

This process of breaking and reconnecting of magnetic fields is now thought to be fundamental in explaining the dynamic and highly variable solar corona and Earth's magnetosphere. The basic process is

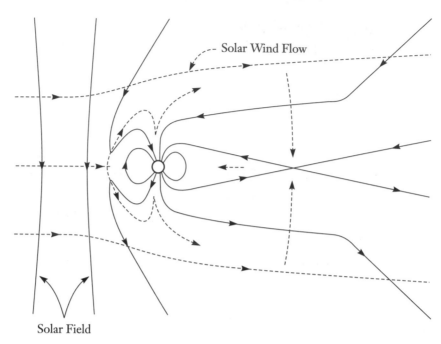

Solar Wind Flow

Solar Field

FIGURE 8.3   Interaction of the magnetized solar wind with the Earth's magnetosphere.

illustrated in Figure 8.4, which shows oppositely-directed magnetic
fields being pushed together, although finding the "pushing" agent is a
major problem in most theories. When the oppositely-directed fields
come in contact they change their connectivity, so that instead of up-
down field lines we end up with left-right field lines. All of this action
occurs in a small reconnection region, a black box within which most of
the interesting and complicated physics happens. The result of this
reconnection process is that jets of energetic plasma shoot out from the
reconnection site; in the case of the Earth's magnetotail (the part of the
Earth's magnetic field on the side distant from the Sun, where it is
blown and narrowed into a long tail by the solar wind), this process
feeds the plasma sheet that is implicated in magnetic substorms.

The reconnection process, which occurs on the dayside (the side of
the Earth facing the Sun), leads to an assortment of dramatic effects in

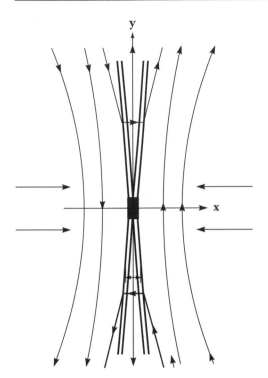

FIGURE 8.4   Reconnection of magnetic fields.

Earth's magnetosphere. First, the field lines of the magnetosphere that are involved in the opening-up process spring back toward the Earth's poles, and they bring some of the solar wind particles back with them. These field lines are then blown back toward the magnetotail, and the hot plasma (after first entering the polar regions and being bounced back out by the increasing strength of the Earth's polar fields) flows back into the tail region.

In the tail, about 100–200 Earth radii back, yet another reconnection process occurs, because at these locations the magnetic fields are again oppositely-directed. In this case, the reconnection process shoots a jet of high-energy plasma back in toward the Earth (and also outward away from the Earth), and this plasma forms the plasma sheet that was shown in Figure 8.2. Particles from this sheet drift in toward Earth and are trapped in the closed magnetic field of the interior parts of the magnetosphere, eventually forming donut-shaped rings of energetic particles known as the Van Allen radiation belts.

The first discovery of the space age was the Van Allen belts, regions of charged particles trapped in the Earth's magnetic field. James A. Van Allen, with the first U.S. satellite, Explorer-1, discovered the radiation belts in 1958 and found that there were at least two distinct parts, an inner and an outer belt. Since then a number of satellites have been launched to study the radiation belts. Among other things, scientists have discovered that there are actually many separate populations in the radiation belts including various electron belts, proton belts, and belts made up of anomalous cosmic rays trapped in the magneto-sphere—which Explorer-1's crude Geiger counter could not distin-guish. Van Allen's two belts are now known to be the two main electron belts.

As the particles bounce around from pole to pole in the magnetic field, they are pushed into the east-west direction by the magnetic field. Positively-charged particles—protons and ions—drift westward around the magnetic equator, and negatively charged particles—elec-trons—drift eastward. The net result is a huge ring current circling the Earth in the radiation belts.

FLARES

In this discussion of space weather we are interested in solar flares mainly for their effects on Earth and on the near-Earth space environ-ment. To this end, there are two main products of these solar instabili-ties relevant to the discussion:

1. There are enormous eruptions of coronal material known as co-ronal mass ejections, which may or may not always be caused by solar flares (this is a controversial point at the moment), and which send clouds of hot, magnetized plasma out into interplanetary space.

2. In a flare, the emission of electromagnetic radiation, from radio

to gamma rays, is greatly increased, and large numbers of high-energy particles are also produced.

These flare products produce a wide variety of effects at Earth: increased particle precipitation into the polar regions, dynamic variations in the Earth's plasma tail leading to geomagnetic storms, large ring currents in the ionosphere leading to ground-level power surges via magnetic induction, high levels of energetic particles impinging on spacecraft, upper atmospheric fluctuations that interfere with communications and other electronic functions such as the Global Positioning System, and many others too numerous to list. In the next sections we explain how these solar emissions change the Earth environment to produce this variety of harmful effects.

Plate XV shows a series of images taken by the TRACE satellite, showing the development of a flare. The first frame shows the preflare state of an active region in the corona, seen in the extreme ultraviolet (EUV) at a temperature of 1 million kelvins. The magnetic fields in the region are generally oriented left-right, and the fine structure seen in the EUV follows this orientation.

But there is also a region of magnetic fields emerging into the preexisting region, and it is oriented up-down in these images. It is the interaction of this new field with the old that causes the instability that we call a flare. This is a classic magnetic field configuration for the production of flares, because it involves fields having opposite directions being pressed together, which is in turn a configuration that is a leading contender for magnetic reconnection.

The reason that reconnection occurs is that energy is stored in magnetic fields, and when they are loaded with hot matter there is additional energy in the structure. This energy would normally be released fairly slowly, but when the magnetic structures become complex and tangled, it may happen that there is a simpler field geometry that represents a lower energy state. Reconnection of the field lines can then allow a rearrangement to the lower-energy configuration, and the difference in energy is released in the process. This extra energy can go into accelerating the hot plasma outward, or heating it to tens of millions of degrees, or both. It is this rapid energy release that we call a flare, and the new observations from the Yohkoh, SOHO, and TRACE satellites are showing the way to a better understanding of these dynamic events in the corona.

CORONAL MASS EJECTIONS

Effects due to magnetic reconnection occur in even more dramatic fashion when the reconnection drives a coronal mass ejection, which carries more hot plasma, stronger magnetic fields, and higher-energy particles than does the typical solar wind. Coronal mass ejections are not easily detected from ground-based instruments, and it was only when coronagraphs were flown in space in the early 1970s that these enormous eruptions were clearly seen. A typical coronal mass ejection spews out 20 billion tons of coronal material at speeds of several hundred miles per second. This is roughly a tenth of the total coronal mass, and the amount of energy released is, in standard U.S. units, up to a million trillion kilowatt-hours.

It is through coronal mass ejections that solar activity is most forcefully felt at the Earth. They are responsible for essentially all of the largest energetic proton events—in which these high-energy particles bombard the Earth and the space environment near the Earth—and they are the primary cause of geomagnetic storms (Plate XVI). One such event, observed by the LASCO coronagraph on the SOHO satellite, is shown in Figure 8.5. When such ejections are directed toward the Earth, they can produce the entire range of geomagnetic effects described earlier in this chapter.

THE UPPER ATMOSPHERE

The influence of solar variability increases with height in our atmosphere, and in space the protective effects become negligible. The

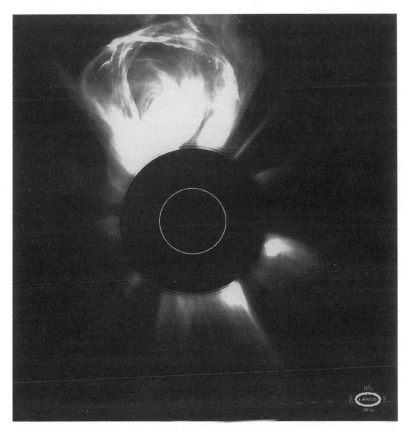

FIGURE 8.5   A coronal mass ejection, as seen by the LASCO C2 coronagraph, 27 February 2000. The size of the Sun's photosphere is indicated by the white circle superposed on the occulting disk that hides the photosphere and the region around it.

Earth's magnetic field continues to provide some protection at high altitudes, but it also increases the hazards by funneling energetic particles toward the poles and by storing energetic particles in radiation belts. Transient "solar storm" events trigger complex short-term disturbances in the magnetosphere which are felt especially strongly in the upper atmosphere and which can be fatal to both instrumentation and living beings in space. The ultraviolet (UV) radiation from the Sun has a strong effect on the upper atmosphere, producing the protective ozone layer, but also reflecting changes in the incoming solar UV intensity by changes in the observed properties of the middle and upper atmosphere.

THE IONOSPHERE

In common parlance, the word stratosphere signifies a sort of ultimate upper extreme. In actuality, the Earth's atmosphere extends a great deal farther, and is classified into regions based largely on the way that the temperature varies with height:

Troposphere	The region from ground level to about 7 miles up; this is where clouds and most weather occur. The temperature generally drops with height, to about −60°F at the top of this region.
Stratosphere	From about 7 to 30 miles up, this region is relatively isolated from the troposphere, so it is usually free of water vapor and dust. Absorption of the incoming solar ultraviolet light by ozone warms the air to about 40°F.
Mesosphere	From 30 to 50 miles up, this is a region character-ized by falling temperatures, down to about −90°F. It is poorly explored because it is too high for air-craft or balloons and too low for satellites, so obser-vation is difficult.
Ionosphere	Also known as the thermosphere, this is the region from 50 miles up. The incoming solar radiation has a large effect on this region, causing temperature to vary between 900°F and 3,000°F; the upper extent of the region fluctuates accordingly, and satellites in low-Earth orbit are affected by the atmospheric drag at times of high solar activity.

The large temperature changes in the ionosphere may seem surpris-ing, especially since the solar wavelengths involved, which are mainly the UV and EUV, are only a small fraction of the total solar output. But

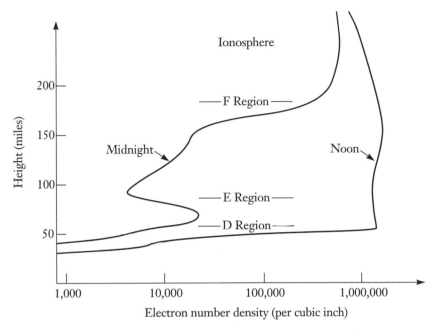

FIGURE 8.6   The number density of electrons in the ionosphere is strongly affected, on a daily basis, by the short-wavelength radiation from the Sun.

the Earth's atmosphere is so rarefied at those altitudes that there just
isn't very much of it, and it doesn't take much energy to change the
temperature by a large amount. The reason it is called ionosphere is
that the incoming solar radiation knocks electrons out of the atoms in
the air, creating positively charged ions and the associated free electrons. Figure 8.6 shows how the number of electrons varies with height
in the ionosphere, with one curve for daytime and another curve for
night. The effect of the Sun is clearly visible between heights of about
50 and 180 miles, with the curve labeled "Noon" showing an increase
of more than a factor of 100 in the number of electrons compared to
the curve labeled "Midnight."

OZONE

Chlorofluorocarbons, or CFCs, are a problem that started out as a solution. They were invented in 1928 by Thomas Midgley, a chemist at

General Motors, who was looking for a refrigerant to replace the toxic and explosive ammonia being used at that time. CFCs seemed ideal: the dangerous chlorine is locked up in CFC, in a nonreactive form that did not seem to break down if it was released into the atmosphere. CFCs were also found to work well as a propellant (for items such as spray deodorants) and as a solvent.

It was only in the 1970s that atmospheric scientists began to investigate where all of the CFCs were going, and whether they were having any effect on our atmosphere. Soon it became clear that the CFCs continue rising high into the stratosphere, where the ultraviolet radiation from the Sun makes them break down and release the chlorine. The chlorine then feeds a complicated chain of chemical reactions that ends up destroying thousands of ozone molecules for every chlorine atom.

Why is destruction of ozone a problem? Ozone high in the atmosphere absorbs much of the incoming short-wavelength ultraviolet radiation from the Sun, thereby protecting us from these damaging rays, which can cause skin cancer and can also damage crops. Ozone is both formed and destroyed by ultraviolet from the Sun. The longer-wavelength, lower-energy ultraviolet is absorbed high in the atmosphere, and it breaks apart oxygen molecules (with 2 atoms of oxygen) into single atoms of oxygen. These bounce around until they hit other oxygen molecules and combine with them to form the 3-atom combination we call ozone. The shorter-wavelength, higher-energy UV penetrates slightly lower in the atmosphere, where it breaks apart the ozone molecules, and gets absorbed in the process.

Normally, there is a balance between the production of ozone and the natural processes that remove it. But the addition of chlorine from CFCs greatly alters this balance, leading to a massive depletion of ozone. This was first noticed in 1985 in measurements over the Antarctic, which showed that nearly half of the normal ozone was missing during Southern Hemisphere spring months. Satellite observations confirmed the existence of this "ozone hole," which continued to increase in size each year.

The mechanism for ozone depletion over the south pole seems to be the special condition in winter. As temperatures fall, the high-latitude vortex of air around the pole becomes stronger and prevents the polar air from mixing with the rest of the Earth's atmosphere. A reservoir of chlorine forms that enhances ozone destruction, and in the spring this ozone-poor air is released to spread over the Southern Hemisphere.

Beyond the particular worries about ozone over the Antarctic, these observations raise the larger question of whether this is an early warning of a more global problem. Global climate changes often show up first at high latitudes, where the effects can be more extreme; these can provide us with advance indication of a spreading problem, if we are willing to heed the warning signs. In this instance, the problem was recognized, and CFC (also called Freon) production has decreased in recent years. However, the residence time for CFCs in the upper atmosphere is decades, so it will be some time before the situation begins to improve.

AFTERWORD

So we end where we started, with the influence of the Sun on the Earth. The effects are seen to be multiform and complex, with potential dangers from literally all directions. We often hear it said that humans are increasingly able to control their environment, but a closer look shows that our culture has also become increasingly dependent on a controlled environment. For instance, farmers are abandoning certain types of less profitable crops that used to be grown in off-seasons because of their ability to control the growth environment. Housing is being built in locations that would have been too risky in the past—on slopes subject to mud slides, in coastal regions subject to storm damage, in desert areas where water supply is an issue. We have contructed and rely on elaborate power distribution networks, and we are becoming increasingly dependent on reliable satellite communications.

Such activities make us more dependent on our ability to retain control over the environment. To the extent that we have been assuming the constancy of the terrestrial and near-Earth climate, we are becoming more, rather than less, dependent on the constancy of the Sun. What we have shown in this book, we hope, is that this dependency requires that we continue our efforts to understand the physics of the Sun—that we develop new and better observing instruments as well as new and better theoretical models.

What is new today is that our planet's fate, as well as our ability to prosper as a human community, is tied not only to the Sun, but also to our understanding of the Sun.

Bibliography

Bahcall, J. N. *Neutrino Astrophysics.* Cambridge, Eng.: Cambridge University Press, 2000.

Bahcall, J. N., F. Calaprice., A. B. McDonald, and Y. Totsuka. "Solar Neutrino Experiments: The Next Generation." *Physics Today,* July 1996, pp. 30–36.

Barrow, J. D. and F. J. Tipler. *The Anthropic Cosmological Principle.* Oxford: Oxford University Press, 1988.

Beatty, J. K., C. C. Petersen, and A. Chaikin. *The New Solar System,* 4th ed. Cambridge, Eng.: Cambridge University Press, 1998.

Bester, A. *The Life and Death of a Satellite.* Boston. Little Brown, 1966.

Budyko, M. I. *Climatic Changes.* Washington, D.C.: American Geophysical Union, 1977.

Chapman, A. "The Transits of Venus." *Endeavour,* 22 (1998): 148–151.

Eddy, J. A. "The Maunder Minimum." *Science,* 192 (1976): 1189.

Fernie, J. D. "Transit, Travels, and Tribulations, I." *American Scientist,* 85 (1997): 120–122.

———"Transit, Travels, and Tribulations, II." *American Scientist,* 85 (1997): 418–421.

———"Transit, Travels, and Tribulations, III." *American Scientist,* 86 (1998): 123–126.

———"Transit, Travels, and Tribulations, IV." *American Scientist,* 86 (1998): 422–425.

Golub, L., and J. M. Pasachoff. *The Solar Corona.* Cambridge, Eng.: Cambridge University Press, 1997.

Gough, D. O., J. W. Leibacher, P. H. Scherrer, and J. Toomre. "Perspectives in Helioseismology." *Science,* 272 (1996): 1281–1283.

Guillermier, P., and S. Koutchmy. *Total Eclipses: Science, Observations, Myths and Legends.* New York: Springer, 1999.

Harrington, P. S. *Eclipse! The What, Where, When, Why, and How Guide to Watching Solar and Lunar Eclipses.* Chichester and New York: John Wiley, 1997.

Hirsh, R. F. *Glimpsing an Invisible Universe.* Cambridge, Eng.: Cambridge University Press, 1983.

Imbrie, J., and J. Z. Imbrie. *Ice Ages: Solving the Mystery,* 2nd. ed. Cambridge, Mass.: Harvard University Press, 1980.

Intergovernmental Panel on Climate Change. *Climate Change: The IPCC Scientific Assessment.* Cambridge, Eng.: Cambridge University Press, 1990.

————*Climate Change 1995: The Science of Climate Change: Contribution of Working Group I to the Second Assessment Report of the IPCC.* Cambridge, Eng.: Cambridge University Press, 1996.

Jager, J. *Climate and Energy Systems.* New York: John Wiley, 1983.

Kaneyuki, K., and K. Scholberg. "Neutrino Oscillations." *American Scientist,* 187 (1999): 222–231.

Lamb, H. H. *Climate, History, and the Modern World,* 2nd ed. London: Routledge, 1995.

————*Climate: Past, Present and Future,* vol. 1. London: Methuen, 1972.

Lang, K. *Earth, Sun, and Sky.* New York: Copernicus Books, 1997.

Lean, J. "The Sun's Variable Radiation and Its Relevance for Earth." *Annual Reviews of Astronomy & Astrophysics,* 35 (1997): 33.

————"Variations in the Sun's Radiative Output." *Reviews of Geophysics,* 29 (1991): 505.

Maor, E. *June 8, 2004: Venus in Transit.* Princeton: Princeton University Press, 2000.

Mariska, J. T. *The Solar Transition Region.* Cambridge, Eng.: Cambridge University Press, 1992.

Menzel, D. H. *Our Sun,* 2nd ed. Cambridge, Mass.: Harvard University Press, 1959.

National Research Council. *Solar Influences on Global Change.* Washington, D.C.: National Academy Press, 1994.

Newton, R. R. *Medieval Chronicles and the Rotation of the Earth.* Baltimore: The Johns Hopkins University Press, 1972.

Noyes, R. W. *The Sun, Our Star.* Cambridge, Mass.: Harvard University Press, 1982.

Pasachoff, J. M. *Peterson Field Guide to the Stars and Planets,* 4th ed. Boston: Houghton Mifflin, 2000.

Payne, C. *Stellar Atmospheres: A Contribution to the Observational Study of High Temperature in the Reversing Layers of Stars.* Cambridge, Mass.: Harvard Observatory Monograph no. 1, 1925.

Phillips, K. J. H. *Guide to the Sun.* Cambridge, Eng.: Cambridge University Press, 1992.

Sagan, C., and G. Mullen. "Earth and Mars: Evolution of Atmospheres and Surface Temperatures." *Science,* 177 (1972): 52.

Sonett, C. P., M. S. Giampapa, and M. S. Matthews. *The Sun in Time.* Tucson: University of Arizona Press, 1991.

Stephenson, F. R. *Historical Eclipses and Earth's Rotation.* Cambridge, Eng.: Cambridge University Press, 1997.

Stevens, William K. *The Change in the Weather: People, Weather, and the Science of Climate.* New York: Delacorte Press, 1999.

Suess, S. T., and B. T. Tsurutani, ed. *From the Sun: Auroras, Magnetic Storms, Solar Flares, Cosmic Rays.* Washington, D.C.: American Geophysical Union, 1998.

Ulrich, R. K. "Solar Neutrinos and Variations in the Solar Luminosity." *Science*, 190 (1975): 619.

Ward, P. D., and D. Brownlee. *Rare Earth: Why Complex Life Is Uncommon in the Universe.* New York: Copernicus Books, 2000.

Wenzel, D. *The Restless Sun.* Washington, D.C.: Smithsonian Institution Press, 1989.

Williams, J. *The Weather Book*, 2nd ed. New York: Vintage Books, 1997.

World Council of Churches. *Climate Change and the Quest for Sustainable Societies.* Geneva: World Council of Churches, 1998.

Glossary

ACE   Advanced Composition Explorer. A NASA-launched satellite located at the Earth-Sun Lagrange point L1, designed to study the element and isotope composition of particles, including those from the solar wind.

ACRIM   Active Cavity Radiance Irradiation Monitor. An experiment onboard the SMM satellite that measured the total radiative output of the Sun.

active region   The portion of the solar surface and atmosphere where strong magnetic fields, sunspots, plages, flares, and other activity are found.

angstrom   A unit of length equal to 10^{-10} meters (symbol Å).

albedo   The ratio of the amount of light reflected back from a surface to the amount of incoming light.

alpha particle   A helium atom nucleus, emitted in radioactive decay of heavy elements.

atom   The smallest unit of a chemical element; when it is subdivided, the parts no longer have the properties of a chemical element.

aurora   Glowing lights visible in the sky, due to excitation of atoms and molecules by downward-flowing high-energy electrons.

Baily's beads   Bright dots of light visible around the rim of the Moon at the beginning and end of a total solar eclipse.

Balmer series   The set of spectral lines resulting from transitions to or from the first excited energy level in the hydrogen atom; compare with the Lyman series.

baryon   A nuclear particle, such as a proton or neutron, subject to the strong nuclear force, made of triplets of quarks.

beta particle   An energetic electron or anti-electron (positron) emitted from a nucleus in radioactive decay; also called beta ray for historical reasons.

black body   A hypothetical object that absorbs all radiation that hits it and emits radiation exactly according to Planck's Law.

blue shift   A change toward shorter wavelengths in an object's spectral feature; when the shift is caused by motion, it results from the velocity at which the object is approaching us.

bolometer   A device for measuring the total amount of energy received from an object.

bolometric luminosity   The brightness of an object, with all wavelengths taken into account.

calorie   A measure of energy in the form of heat; originally the amount of heat needed to raise the temperature of 1 gram of water by 1°C.

carbon cycle   A chain of nuclear reactions, involving carbon as a catalyst in some of its stages, that fuses four hydrogen nuclei into one helium nucleus, with a resulting release of energy. This cycle is important in stars hotter than the Sun.

Cassegrain   A type of telescope in which light focused by a primary mirror is intercepted short of the focal point and reflected by a secondary mirror through a hole in the center of the primary mirror; compare with Gregorian.

catalyst   A substance that participates in a reaction, but that is left over in its original form at the end.

celestial equator   The intersection of the celestial sphere with a plane passing through the Earth's equator.

celestial sphere   An imaginary sphere with a large radius, centered on the Earth, to which the stars appear to be fastened.

Cenozoic   The period beginning about 70 million years ago, characterized by the appearance of mammals.

chromatic aberration   A defect of an optical system in which different colors are focused at different points.

chromosphere   The part of the solar (or stellar) atmosphere between the photosphere and the corona.

convection   The transfer of heat by motion of the hotter parts of a gas or liquid.

corona   A portion of the solar atmosphere characterized by temperatures above 1 million degrees Celsius.

coronal hole   A relatively dark region of the corona having lower density, due to magnetic field lines open to interplanetary space.

cosmic rays   High-energy particles or nuclei traveling through space.

cosmology The study of the Universe as a whole.

density Mass divided by volume.

deuterium An isotope of hydrogen, containing one proton and one neutron in the nucleus.

differential rotation Rotation of a body in which different parts (typically different latitudes) have different angular velocities, and therefore different rotation periods.

diffraction A spreading out of light after it passes an obstacle.

diffraction grating An object having a closely spaced series of lines that cause different wavelengths to be spread out in space, thereby providing a spectrum of the light.

dispersion Of light, the effect that different wavelengths are bent by different amounts in passing from one material to another. The effect is caused by differences in the speed of light for different wavelengths in a medium.

D lines Usually refers to a pair of lines from sodium that appear in the yellow part of the spectrum.

dynamo A device for generating strengthened magnetic fields via the motions of a conducting fluid.

dynamo theory An explanation of magnetic field generation via a mathematical model.

eccentricity For a planetary orbit, a measure of the flatness; for an ellipse, it is given by the distance between the two foci divided by the length of the major axis.

eclipse The passage of all or part of one astronomical body into the shadow of another.

ecliptic The plane of the Earth's orbit; also, the apparent annual path traced out by the Sun against the celestial sphere as seen from Earth.

electron A particle with one electric charge, 1/1,830 the mass of a proton, that is not affected by the strong nuclear force.

electron volt The energy gained by an electron when accelerated by 1 volt.

element An atom characterized by a specified number of protons in its nucleus.

emission line Sharp energy peaks in a spectrum, caused by downward transitions from one discrete energy level to another.

energy level A quantum state of a system, such as an atom, having a well-defined energy.

ephemeris A listing of astronomical positions and other data that change with time; from the same root as "ephemeral."

epoch A subdivision of a geological period, characterized by a particular dominant characteristic.

era A major division of geological time, composed of a number of periods.

erg A unit of energy, corresponding to the work done by a force of 1 dyne (the force needed to accelerate 1 gram by 1 centimeter/second^2) acting over a distance of 1 centimeter.

excitation The raising of electrons to higher energy states in an atom.

filament A dark, elongated feature of the solar surface seen in H-alpha; a prominence when projected against the sky.

flare An extremely rapid brightening of a small volume of the solar atmosphere, with portions reaching a temperature of more than 10 million degrees Celsius.

flash spectrum The solar chromospheric spectrum seen in the few seconds just before or just after totality at a solar eclipse.

flux The amount of something, such as energy, passing through a surface per unit time (per second, for example).

forcing (external) A forced adjustment of the internal properties of a system via a quasi–steady-state change in boundary conditions.

Fraunhofer lines The absorption lines of the solar or stellar spectrum.

gamma rays Electromagnetic radiation shorter than about 0.01 nanometer (0.1 angstrom).

geocentric Earth-centered.

granulation On the Sun, convection cells about 1 arc second across.

grating A surface ruled with closely spaced lines that, through diffraction, breaks up light into its spectrum.

grazing incidence Striking at a small angle.

greenhouse effect The warming of a planet's surface due to trapping of infrared radiation by the atmosphere.

Gregorian A type of telescope in which light focused by a primary mirror passes through its focal point and is then reflected by a secondary mirror, forming an image through a hole in the center of the primary mirror; compare with the Cassegrain.

ground level The lowest energy level of a quantum-mechanical system; also called ground state.

H line The spectral line of ionized calcium at 396.8 nanometers (3,968 angstroms).

H-alpha The first line of the Balmer series of hydrogen at 656.3 nanometers (6,563 angstroms).

heliocentric Sun-centered.

helium flash The rapid onset of fusion of helium into carbon through the triple-alpha process that takes place in most red giant stars.

hertz A measure of frequency with units of 1 over seconds, that is, per second.

H-R diagram Hertzsprung-Russell diagram; a graph of temperature (or the equivalent) versus luminosity (or the equivalent) for a group of stars.

inclination Of a planetary orbit, the angle of the plane of the orbit with respect to the ecliptic.

insolation The rate of solar energy hitting an area on Earth.

interference The property of waves that waves in phase can add (constructive interference) and those out of phase can cancel (destructive interference); for light, interference produces alternating light and dark bands.

interferometer A device that uses interference to measure properties of a source, such as position or structure.

ion An atom that is missing one or more electrons.

ionosphere The highest region of the Earth's atmosphere, from 50 miles up; also called the thermosphere.

isotope Of an element, one having the same number of protons, but a different number of neutrons.

joule A unit of energy, equal to $1 \text{ kg} \cdot \text{m}^2/\text{s}^2$; used instead of erg in the International System of Units; 1 joule = 10,000,000 ergs.

Julian day The number of days since noon on 1 January 713 B.C.E. Noon of 1 January 2000 was Julian day 2,451,545.

K line The spectral line of ionized calcium at 393.3 nanometers (3,933 angstroms).

Lagrange points In the restricted three-body problem, there are five equipotential points; two are marginally stable and three are marginally unstable.

L1 The Lagrange point between the Earth and the Sun; located about 1 million miles from Earth.

limb The edge of an extended object, such as a star or planet.

limb darkening The decreasing brightness of the disk of the Sun as one looks from the center toward the limb.

luminosity   The total amount of energy given off by an object per unit of time.

lunar eclipse   The passage of the Moon into the Earth's shadow.

Lyman alpha   The spectral line at 121.6 nanometers (1,216 angstroms) that corresponds to a transition between the two lowest energy levels of the hydrogen atom.

Lyman series   The series of spectral lines in the ultraviolet corresponding to transitions to or from the ground state of the hydrogen atom.

Mach number   The number of times the speed of an object is greater than the local speed of sound.

magnetic field   A property of space that can exert magnetic forces on objects.

magnetic field lines   Directions mapping out the magnetic forces in space; the degree of packing of field lines indicates the relative strength of the field.

magnetosphere   A region of magnetic field around a planet.

main sequence   A narrow band in the H-R diagram in which stars fall during the hydrogen-burning phase of their lifetimes.

mass   A quantity indicating the amount of force needed to accelerate an object (inertial mass) or the strength of the gravitational force produced by an object (gravitational mass); by Einstein's principle of equivalence in the general theory of relativity, the two are equal.

Maunder minimum   The period 1645–1715, when very few sunspots were visible.

mesosphere   A middle layer of the Earth's atmosphere above the stratosphere and below the ionosphere.

Milankovitch periodicity   The periodic variation in the Earth's climate due to variations in orbital parameters (obliquity, precession, and eccentricity).

nebula   An interstellar region of gas or dust.

neutrino   A neutral elementary particle interacting weakly with matter, but carrying mass, energy, momentum, and spin.

neutron   A massive, neutral elementary particle, interacting via the strong nuclear force.

nuclear force   The strong force, one of the four known fundamental forces.

nucleosynthesis   The formation of the elements.

nucleus   Of an atom, the core, which has positive charge, most of the mass, and occupies a small part of the volume.

objective   The principal lens or mirror of an optical system.

obliquity A measure of the tilt of the Earth, given by the angle between the Earth's equatorial plane and the plane of the ecliptic.

occultation The hiding of one astronomical body by another.

optical The visible part of the spectrum, 390–660 nanometers (3,900–6,600 angstroms).

ozone A molecule containing 3 oxygen atoms, O_3.

ozone layer. A region of the Earth's upper stratosphere and lower mesosphere where ozone absorbs ultraviolet radiation.

parallax. The change in the apparent position of an object with respect to more distant objects that results from a changing perspective.

penumbra For an eclipse, the part of the shadow from which the Sun is only partially occulted; of a sunspot, the outer region, not as dark as the umbra.

period The basic unit of geological time, during which a standard type of rock system formed; a period typically comprises several epochs, and several periods are typically combined to constitute an era.

photosphere The region of a star from which most of its light is radiated.

plage The part of an active region that appears bright when viewed in H-alpha.

Planck's law The relation between wavelength of radiation and intensity at that wavelength for a given temperature. For each temperature, Planck's law defines a peaked curve relating wavelength and intensity.

plasma A hot gas in which a substantial fraction of the atoms are ionized.

plumes Thin vertical structures in the solar corona near the poles.

pressure Force per unit area.

prime focus The location at which the main optical element of a telescope would form an image.

prominence Dense solar gas at chromospheric temperatures protruding above the limb.

proton Massive elementary particle with unit positive charge, interacting via the strong force.

precession The slow variation in position of the celestial sphere, due to the wobble of the Earth's rotation axis.

pulsar A source of regular pulses of radio waves with intervals ranging from milliseconds to seconds. Discovered by Jocelyn Bell in 1967 and since realized to be a rotating neutron star, a collapsed massive star, with a narrow beam sweeping across us each time the star rotates.

quasar   Discovered as a "quasi-stellar radio source," an object, corresponding to a strong radio source, that appeared optically almost as a point like a star but often with some spatial structure barely visible. Quasars, some of which are not radio sources, have huge velocities of recession and are therefore among the most distant objects in our expanding universe. They are now considered to be extremely active galactic nuclei, presumably fueled by black holes containing millions or billions as much mass as the Sun.

quark   One of the fundamental constituents of matter. Quarks come in six types—up, down, strange, charmed, top, and bottom (or truth and beauty); sets of three quarks make up particles like protons and neutrons. The family containing the up and down quarks as well as electrons and electron neutrinos make up normal matter.

Quaternary   The latter part of the Cenozoic era, beginning about 2.4 million years ago and including the present.

quiet Sun   A collection of solar phenomena showing little variation with the solar cycle.

radiation   Electromagnetic energy; sometimes also energetic particles.

radiation belts   Toroidal bands of charged particles surrounding the Earth and other planets.

red giant   A stage in a star's life after it has lived the main part of its lifetime (which corresponds to its temperature and luminosity lying on the main sequence of the Hertzsprung-Russell diagram). Red giants are large but cool stars. The Sun will become a red giant in about 5 billion years.

reflecting telescope   A type of telescope that uses mirrors to form the image.

refracting telescope   A type of telescope that uses lenses to form the image.

refraction   The bending of the direction of an electromagnetic wave at the interface between two media or in travel through a variable medium.

resolution   The ability of an optical system to distinguish detail.

revolution   The orbiting of one body around another.

rotation   Spin on an axis.

seeing   The steadiness of the Earth's atmosphere as it affects the image quality of a ground-based telescope.

seismology   The study of waves propagating through a body, often to determine the internal properties of the body.

shock wave   A front marked by a sharp change in pressure, caused by an object moving through a medium faster than the local sound speed.

sidereal   With respect to the stars.

SMM Solar Maximum Mission. A satellite designed to study solar flares, launched in 1980 and operational until the end of 1989.

SOHO A joint ESA/NASA solar satellite, launched 2 December 1995, and positioned at the Earth-Sun Lagrange point L1.

solar atmosphere The photosphere, chromosphere, transition region, and corona.

solar constant The total amount of energy from the Sun hitting the Earth, measured at the top of the atmosphere.

solar cycle The 11-year variation in the number of sunspots on the Sun; the full magnetic cycle is 22 years.

solar dynamo The periodic generation and destruction of magnetic fields in the Sun.

solar flare An explosive release of energy on the Sun.

solar wind An outflow of particles from the Sun representing the expansion of the solar corona.

special theory of relativity Einstein's 1905 theory that deals with moving objects; it led to the equivalence of mass and energy in the relation $E = mc^2$ and holds that the speed of light cannot be exceeded by moving objects. It is a theory of invariances in transforming from one unaccelerated frame to another.

spectral line A narrow range of wavelengths within which the intensity of a spectrum is sharply different from that of its surroundings.

spectrograph A device to make and record a spectrum.

spectrometer A device to make and record a spectrum electronically.

spectroscope A device to make and look at a spectrum.

spectroscopy The analysis of spectra.

spectrum A display of electromagnetic radiation spread out by wavelength or frequency.

spicule A small jet of gas seen at the limb of the Sun, about 1,000 kilometers across and 10,000 kilometers high with a lifetime of about 15 minutes.

strong force The nuclear force, the strongest of the four known forces of nature.

subtend To take up an angle in your field of view; for instance, the Sun subtends about 1/2 degree as seen from the Earth.

sunspot A region of the solar surface that is dark and relatively cool, having a strong magnetic field.

sunspot cycle The 11-year periodic variation in the number of sunspots visible on the Sun.

supergranulation   Convection cells on the solar surface about 20,000 kilometers across and roughly polygonal in shape.

syzygy   An alignment of three or more celestial bodies.

thermosphere   The uppermost layer of the Earth's atmosphere; same as the ionosphere.

TRACE   A solar satellite in the NASA Small Explorer program, launched on 2 April 1998.

transit   The passage of one celestial body in front of another.

transition region   The thin region of the solar atmosphere between the chromosphere and the corona.

triple-alpha process   A chain of fusion reactions in which 3 helium nuclei (alpha particles) combine to form a carbon nucleus.

troposphere   The lowest level of the Earth's atmosphere, in which most weather takes place.

ultraviolet   The region of the spectrum from 10–400 nanometers (100–4,000 angstroms).

umbra   Of a sunspot, the dark central region; of an eclipse shadow, the darkest part, from which the solar disk cannot be seen at all.

Van Allen belts   Toroidal regions in the Earth's magnetosphere, in which high-energy particles are trapped.

visible light   Light to which the eye is sensitive, 390–660 nanometers (3,900–6,600 angstroms).

wavelength   The distance over which a wave travels during one complete oscillation.

white dwarf   A collapsed star with a mass approximating that of the Sun in a volume the size of the Earth.

white light   All the light of the visible spectrum together.

x-rays   Electromagnetic radiation between 0.01 and 10 nanometers (.01 and 100 angstroms).

year   The period of revolution of a planet around its central star; particularly, the Earth's period of revolution around the Sun.

Yohkoh   A solar satellite of the Japanese space agency, with U.S. and U.K. participation, launched on 30 August 1991; originally called Solar-A.

Zeeman effect   The splitting of certain spectral lines in the presence of a magnetic field.

Acknowledgments

LG would like to acknowledge helpful discussions with many friends and colleagues, especially: Spiro Antiochos, Jay Bookbinder, Ed De-Luca, Robert Rosner, and Harry Warren. Much of the work described in this book was supported by the National Aeronautics and Space Administration, and he thanks Dave Bohlin, Bill Wagner, and George Withbroe for their help and support. He is especially grateful to his wife, Anne Davenport, for helping to make this book much better than it otherwise would have been.

JMP would like to thank the students at Williams College who read portions of the manuscript and provided helpful comments: Misa Cowee, Alexandru Ene, Sara Kate May, Mark Kirby, and Kevin Russell; and the students in his solar-physics seminar: Chris Foxwell, Gabe Brammer, Rebecca Cover, Ken Dennison, Rossen Djagalov, Joel Iams, Duane Lee, Dan Seaton, Joey Shapiro, Darik Vélez, and Mike Gioiello. We thank Geoff Reeves for material about the Van Allen belts. Jay Pasachoff's work on solar astronomy in recent years has been supported by the National Science Foundation, by NASA, and by the Committee for Research and Exploration of the National Geographic Society. He thanks his wife, Naomi, for her collaboration in academic and nonacademic ways.

In this year of the 100th anniversary of Donald Menzel's birth, we want to celebrate his memory and to thank Florence Menzel and Bob and Harriet Noyes for their long-term friendships. We both thank our editors, Michael Fisher and Nancy Clemente of Harvard University Press, and the two (anonymous) reviewers selected by HUP, who provided many good suggestions for improving this book. We thank Nancy Kutner for her excellent index.

Many friends and colleagues have contributed by reading portions of the manuscript, by answering questions, and by providing material. We especially thank Angélique Fiedler, Hoyt Rogers, Becca McMullen, and Michele Sprengnether for constructive criticisms of several chapters. These friends have all provided valuable suggestions and advice, but of course any remaining errors in the text are entirely our responsibility.

Illustration Credits

Figures

LaPalma by the Royal Swedish Academy of Sciences in the Spanish Observatorio del Roque de los Muchachos of the Instituto de Astrofisica de Canarius. Used with the permission of Mats Löfdahl.

3.10. SOHO/SOI-MDI.

3.11. J. M. Pasachoff.

3.12. Observatoire de Paris, Section de Meudon.

3.13. Mees Solar Observatory, University of Hawaii. Used with the permission of B. LaBonte.

3.14. J. M. Pasachoff.

3.15. NASA/TRACE.

3.16. National Solar Observatory, Sacramento Peak.

3.17. J. M. Pasachoff.

3.18. From Alan Title, *Selected Spectroheliograms;* observer R. W. Noyes.

3.19. J. M. Pasachoff.

3.20. J. M. Pasachoff.

3.21. Big Bear Solar Observatory, New Jersey Institute of Technology, used with the permission of Haimin Wang.

3.22. W. Livingston, National Solar Observatory.

3.23. NRL and NASA.

3.24. A. Daw, SAO, used with the permission of L. Golub.

4.3. Institute for Cosmic Ray Research (ICRR), University of Tokyo.

4.4. SOHO/SOI-MDI.

5.3. F. Espenak, NASA/GSFC.

5.4. J. M. Pasachoff.

5.5. J. M. Pasachoff.

5.6. J. M. Pasachoff.

5.7. High Altitude Observatory/NCAR.

5.8. Jay M. Pasachoff, Stephan Martin, and Daniel B. Seaton, Williams College Expedition.

5.9. F. Espenak, NASA/GSFC.

5.10. F. Espenak, NASA/GSFC.

5.11. SOHO EIT/NASA Goddard Space Flight Center; eclipse/Jay M. Pasachoff, Stephan Martin, and Daniel B. Seaton, Williams College Expedition; SOHO LASCO/U.S. Naval Research Laboratory and NASA.

5.12. Mees Solar Observatory, University of Hawaii. Used with the permission of B. LaBonte.

5.14. J. Raymond, SAO.

6.1. LASCO SOHO Consortium.

6.2. NASA/TRACE.

6.3. G. Nystrom, SAO.

6.4. NASA/TRACE, used with the permission of Joseph Burt.

6.5. G. Withbroe, NASA.

6.6. U. C. Berkeley.

6.7. Lockheed Martin Missiles & Space, and Southwest Research Institute.

6.8. Photo courtesy of NASA/Goddard Space Flight Center. The Johns Hopkins Applied Physics Laboratory.

6.9. SAO and ISAS.

6.10. Lockheed Martin and ISAS.

6.11. NASA/JPL.

8.5. LASCO SOHO Consortium.

Plates

IA. NASA.

IB. Used with the permission of Robin Ciardullo (Pennsylvania State University).

IC. Used with the permission of Robin Ciardullo (Pennsylvania State University).

ID. Used with the permission of Bruce Balick (University of Washington) and Arsen Hajian (U.S. Naval Observatory, http://aries.usno.navy.mil/ad/pne).

IE. Used with the permission of Bruce Balick (University of Washington).

IF. Used with the permission of Bruce Balick (University of Washington).

II. The Global Oscillations Network Group (GONG) project, managed by the National Solar Observatory, a Division of the National Optical Astronomy Observatories, which is operated by AURA, Inc., under a cooperative agreement with the National Science Foundation. Used with the permission of John Leibacher.

III. Southwest Research Institute.

IV. National Solar Observatory, Kitt Peak.

V. SOHO/EIT Consortium.

VI. NASA/TRACE.

VII. Copyright © Wendy Carlos, J. M. Pasachoff, and SOHO/EIT team, NASA/GSFC.

VIII. NASA/TRACE.

IX. NASA/TRACE.

XA. SOHO/SOI-MDI.

XB. SOHO/SOI-MDI.

XI. SOHO/SOI-MDI.

XII. NASA.

XIII. ISAS.

XIV. G. E. Parks (University of Washington) and the UVI Team, POLAR/NASA.

XV. NASA/TRACE.

XVI. NASA/TRACE.

Index